Mathematics Beyond the Calculus

Electrical and Electronic Engineering Design Series
Electric Circuits Analysis and Design

Electronic Circuit Design with Bipolar and MOS Transistors

CMOS Circuit Design Analog, Digital, IC Layout

Digital Design Logic, Memory, Computers

Analog Filter Design

Error Correction Code Design

Computer Science Design Series
Programming with MFC & Visual C++

Mathematics
Arithmetic – Integers, Fractions, Decimals

Algebra – A Clear Presentation

Mathematics Beyond the Calculus

Mathematics Beyond the Calculus

Nicholas L. Pappas, Ph.D.

© **2018 Nicholas L. Pappas, Ph.D.**

All Rights Reserved Worldwide.
Except as permitted under the Copyright Act of 1976, no part of this book may be reproduced in whole or in part in any manner. Not in any form or by any electronic or mechanical means, nor stored in a retrieval system, or, transmitted, in any form or by any means, electronic, mechanical, photocopying, recording, or otherwise, without the express permission of Nicholas L. Pappas, Ph.D.
ISBN 13: 978 1719 440 547   ISBN 10: 1719 440 549

A Message about this Text: The subject is essentially endless. The purpose here is to say enough about the subject so that you, the reader can pursue what interests you.

A Message from the Author: I have worked continuously in the electronics industry since 1950 except for 11 semesters teaching at San Jose State University (Professor and Chair Computer Engineering 1988-1993). There I discovered my talent for teaching such as it may be. After War2 I attended Lehigh University, and then transferred to Stanford where I earned the MS degree and, while working at HP in the early 1950's, the Ph.D. EE degree. (Somehow I did not get the word and formally apply for the BS degree.) Hardware design has been my principal activity. I learned enough about assembly language, Forth, C and C++ to design the software I needed for my projects. My current activity is designing integrated circuits.

Preface

This text is about solving various types of equations using practical mathematical methods. Only the essentials of each topic are discussed.

This is *not* about proving theorems, taking limits, or other matters important to mathematicians.

> "However, the emphasis should be somewhat more on how to do the mathematics quickly and easily, and what formulas are true, rather than the mathematicians' interest in methods of rigorous proof." Richard Feynman

Concepts from Linear Algebra - the determinant, the finite matrix, the eigenvalue – are presented without the distractions of mathematical rigor. You learn solution methods that do not involve guesses. Methods you implement in a straightforward manner.

The operational calculus can be traced back to Oliver Heaviside[1]. Though many scientists preceded Heaviside in introducing operational methods, the systematic use of operational methods in physical problems was stimulated only by Heaviside's work. The methods he created are undoubtedly among the most important ever created.

Heaviside was criticized for his lack of mathematical rigor. Yet his numerous mathematical and physical methods and results proved to be correct when mathematical rigor was incorporated.

The Laplace Transform, a basis for a modern day operational calculus, is a straightforward technique for solving ordinary, partial differential, and, with a few complications, difference equations and a type of integral equation. On the other hand the Z transform solves difference equations without complications. And, Heaviside's differential operator $D = d/dt$ augments the transform methods.

The Laplace Transform transforms equations in one real variable domain, usually the time t domain, to a complex variable domain where the problem at hand is essentially solved. The inverse transform from the

---

[1] O. Heaviside, *Electrical Papers*, vols 1, 2, Macmillan 1892,
   Electromagnetic Theory, vols 1, 2, 3, 1893-1912, reissued by Benn 1922

Mathematics beyond the Calculus

complex variable domain to the real variable domain completes the solution. Understanding the inverse transform requires knowledge of the theory of functions of complex variables.

Our main interest in functions of a complex variable is integration, because integration of the complicated integrals of inverse transforms is amazingly simplified.

The methods of the differential and integral calculus are extended to complex numbers and functions of complex variables. The results produce tremendous analytic methods.

We show how ordinary differential equations. systems of ordinary differential equations, partial differential equations, and difference equations are readily solved by transform and/or differential operational methods. We show that each type of equation is solved in essentially the same way.

We just define the Fourier Series, and show how to create Fourier series representing waveforms.

Integral Equations – This is a hugh subject, which we limit to how the Laplace transform solves integral equations that include the convolution integral.

Galois Finite Fields $GF(2^m)$ are not used to solve equations per se. They are used to implement functions such as error correcting codes, speech recognition, phase array antennas, and Doppler radar.[0] Functions NOT implemented here.

---

[0] Schroeder, M. R., *Number theory in Science and Communication*, Chapter 25, ISBN 3540 620 060

Contents

1. Determinants 1
    1.1 Solution of Equations 1
    1.2 Properties of Determinants 2
    1.3 Evaluation of Determinants 3
    1.4 Minors and Cofactors 4
    1.5 Expansion of Determinants 4
    Cramer's Rule 5
    Problems 6

2. Finite Matrices 7
    2.1 The (7,4) Decoder 7
    2.2 Definition 10
    2.3 Matrix Multiplication 11
    2.4 Matrix Sums and Differences 13
    2.5 The Adjoint Matrix 13
    2.6 Inverse of a Square Matrix 14
    2.7 Inverse Reversal Rule 15
    2.8 Cancellation of Common Factors 15
    2.9 Complex Elements 15
    2.10 Transpose of a Matrix 16
    2.11 Scalar Multiplication 16
    2.12 Linear Dependence 16
    2.13 Rank of a Matrix 16
    2.14 Vandermonde Matrices 17
    Problems 18

3. Eigenvalues and Eigenvectors 19

4. Laplace Transform 21
    4.1 Laplace Transform 21
    4.2 Prologue 22
    4.3 General Transforms 23
    4.4 Specific Transforms 26
    4.5 Partial Fractions 29
    4.6 Periodic Functions 33
    4.7 The Gamma Function 34
    4.8 More Properties of Transforms 35
    Tables of Transforms 41

5. Functions of a Complex Variable 44
- 5.1 Complex Numbers 44
- 5.2 Analytic Functions 48
- 5.3 Integration 50
  - Closed contours 50
  - Line integrals 51
  - Cauchy's Theorem 52
  - Cauchy's Integral formula 54
  - Singularities 55
  - Laurent Power Series 55
  - Residues 57
  - Higher order poles 58
  - Residue Theorem 59
  - Evaluation of integrals 60

6. Inverse Laplace Transform 62

7. Ordinary Differential Equations 65
- 7.1 Solution by Laplace Transform 65
- 7.2 Solution by Differential Operator 68
  - Complementary solution y_c 68
  - Particular solution y_p 70
  - (a) Reduction of order method 70
  - (b) Undetermined coefficients method 71

8. Systems of Ordinary Differential Equations 74
- 8.1 Solution by Differential Operator 74
- 8.2 Solution by Laplace Transform 77

9. Partial Differential Equations 78
- 9.1 Solution by Separation of Variables 78
  - *Initial and Boundary conditions* 80
- 9.2 Solution by Laplace Transform 81

10 Fourier Series 85

Contents

11 The Z Transform 89
- 11.1 Z Transform Defined 89
- 11.2 General Z Transforms 91
- 11.3 Specific Z Transforms 93
- 11.4 Inverse Z Transforms 95
  - Tables of General and Specific Transforms 98

12 Difference Equations 100
- 11.1 Elementary Sequences 100
- 11.2 Solution by Z Transform 101

13 Integral Equations 103

14 Galois Finite Fields $GF(2^m)$ 106
- 14.1 Polynomial Operations 110
- 14.2 Irreducible Polynomials 113
- 14.3 Minimal Polynomials 114
- 14.4 Primitive Roots and Primitive Polynomials 116
- 14.5 Zech Logarithms 117
- 14.6 Constructing a Systematic G Matrix Generator $g(x)$ ... 118
- 14.7 Systematic G and H Matrices are Related 119

Appendix 121
- A1 Galois Field Equations 121
- A1.1 $GF(2^3)$ 122
- A1.2 $GF(2^4)$ 124
- A1.3 $GF(2^5)$ 127
- A1.4 $GF(2^6)$ 131
  - *Galois Algebra Review* 136

Answers to the Problems 138

Index 142

1 Determinants

Determinants were invented to simplify, to make practical, the solution of systems of linear algebraic equations with n unknowns.

1.1 Solution of Equations

Determinants arise in the study of linear systems. For example consider two linear equations with two unknowns x_1, x_2 and constants a, b, c.

(1a) $\quad a_1 x_1 + b_1 x_2 = c_1$

(1b) $\quad a_2 x_1 + b_2 x_2 = c_2$

The process of elimination produces a solution.

(2a) $\quad x_1 = \dfrac{c_1 b_2 - c_2 b_1}{a_1 b_2 - a_2 b_1}$ (2b) $\quad x_2 = \dfrac{c_2 a_1 - c_1 a_2}{a_1 b_2 - a_2 b_1}$ where $a_1 b_2 - a_2 b_1 \neq 0$

Define *determinant* symbols that represent algebraic expressions in anticipation of solving more complex sets of equations.

(2d) $\quad \begin{vmatrix} a_1 & b_1 \\ a_2 & b_2 \end{vmatrix} = a_1 b_2 - a_2 b_1$

(2e) $\quad \begin{vmatrix} c_1 & b_1 \\ c_2 & b_2 \end{vmatrix} = c_1 b_2 - c_2 b_1$

(2f) $\quad \begin{vmatrix} a_1 & c_1 \\ a_2 & c_2 \end{vmatrix} = a_1 c_2 - a_2 c_1$

Then according to equations 2 the solutions written with determinants are as follows (Cramer's rule, page 5). In the numerators observe that for x_1 the c's replace the a's and for x_2 the c's replace the b's.

(3a) $\quad x_1 = \dfrac{\begin{vmatrix} c_1 & b_1 \\ c_2 & b_2 \end{vmatrix}}{\begin{vmatrix} a_1 & b_1 \\ a_2 & b_2 \end{vmatrix}}$ (3b) $\quad x_2 = \dfrac{\begin{vmatrix} a_1 & c_1 \\ a_2 & c_2 \end{vmatrix}}{\begin{vmatrix} a_1 & b_1 \\ a_2 & b_2 \end{vmatrix}}$

Mathematics beyond the Calculus

1.2 Properties of Determinants

1 If two columns (or rows) of a determinant are exchanged the sign of the determinant is changed.

2 If every element in a column (or a row) of a determinant is multiplied by the number m, the value of the determinant is multiplied by m.

3 The value of a determinant is not changed if to every element in a column (or a row) is added the elements of any other column (or a row) each multiplied by number m. For example

$$(4) \quad \begin{vmatrix} a_1 & b_1 & c_1 \\ a_2 & b_2 & c_2 \\ a_3 & b_3 & c_3 \end{vmatrix} = \begin{vmatrix} a_1 & b_1 + mc_1 & c_1 \\ a_2 & b_2 + mc_2 & c_2 \\ a_3 & b_3 + mc_3 & c_3 \end{vmatrix}$$

4 The expansion of a determinant of order n has n! terms.

5 The value of a determinant is not changed when corresponding rows and columns are exchanged. for example determinants 5a and 5b have the same value when a, b, c are expressed as numbers or symbols. For example

$$(5a) \quad \begin{vmatrix} a_1 & b_1 & c_1 \\ a_2 & b_2 & c_2 \\ a_3 & b_3 & c_3 \end{vmatrix} \qquad (5b) \quad \begin{vmatrix} a_1 & a_2 & a_3 \\ b_1 & b_2 & b_3 \\ c_1 & c_2 & c_3 \end{vmatrix}$$

6 If every element in a column (or a row) of a determinant is zero the value of the determinant is zero.

7 If 2 columns (or rows) of a determinant are identical the value of the determinant is zero.

8 If every element in a column (or a row) of a determinant is expressed as the sum of two or more terms, the determinant may be expressed as the sum of two or more determinants.

1 Determinants

1.3 Evaluation of Determinants

Examples - Manipulate rows and columns using the properties.

(6) $detA = \begin{vmatrix} 1 & 2 & 3 \\ 2 & 4 & 1 \\ 1 & 3 & 0 \end{vmatrix} \xrightarrow[add - r_1 \; to \; r_3]{add - 2r_1 \; to \; r_2} \rightarrow detA = \begin{vmatrix} 1 & 2 & 3 \\ 0 & 0 & -5 \\ 0 & 1 & -3 \end{vmatrix}$

$\xrightarrow[add - 3c_1 \; to \; c_3]{add - 2c_1 \; to \; c_2} \rightarrow detA = \begin{vmatrix} 1 & 0 & 0 \\ 0 & 0 & -5 \\ 0 & 1 & -3 \end{vmatrix}$

$\rightarrow add \; 3c_2 \; to \; c_3 \rightarrow detA = \begin{vmatrix} 1 & 0 & 0 \\ 0 & 0 & -5 \\ 0 & 1 & 0 \end{vmatrix}$

$\rightarrow exch \; r_2 \; \& \; r_3 \rightarrow detA = -1 \begin{vmatrix} 1 & 0 & 0 \\ 0 & 1 & 0 \\ 0 & 0 & -5 \end{vmatrix} = (-1)(-5) \begin{vmatrix} 1 & 0 & 0 \\ 0 & 1 & 0 \\ 0 & 0 & 1 \end{vmatrix} = 5 \; detI = 5$

Manipulate rows. Add ± integer and fractional multiples of rows.

(7) $detA = \begin{vmatrix} 1 & 1 & 1 \\ 2 & -3 & 2 \\ -1 & -3 & -2 \end{vmatrix} \xrightarrow[add \; r_1 \; to \; r_3]{add - 2r_1 \; to \; r_2} \rightarrow \begin{vmatrix} 1 & 1 & 1 \\ 0 & -5 & 0 \\ 0 & -2 & -1 \end{vmatrix}$

$\rightarrow add - 3r_3 \; to \; r_2 \rightarrow \begin{vmatrix} 1 & 1 & 1 \\ 0 & 1 & 3 \\ 0 & -2 & -1 \end{vmatrix} \xrightarrow[add \; 2r_2 \; to \; r_3]{add - r_2 \; to \; r_1} \rightarrow \begin{vmatrix} 1 & 0 & -2 \\ 0 & 1 & 3 \\ 0 & 0 & 5 \end{vmatrix}$

$\rightarrow add \; \tfrac{2}{5} r_3 \; to \; r_1 \rightarrow \begin{vmatrix} 1 & 0 & 0 \\ 0 & 1 & 3 \\ 0 & 0 & 5 \end{vmatrix} \rightarrow add - \tfrac{3}{5} r_3 \; to \; r_2 \rightarrow \begin{vmatrix} 1 & 0 & 0 \\ 0 & 1 & 0 \\ 0 & 0 & 5 \end{vmatrix}$

$\rightarrow add - \tfrac{4}{5} r_3 \; to \; r_3 \rightarrow \begin{vmatrix} 1 & 0 & 0 \\ 0 & 1 & 0 \\ 0 & 0 & 1 \end{vmatrix} = I_3$

Manipulate columns. Add ± integer multiples of columns.

(8) $detA = \begin{vmatrix} 4 & -4 & -4 \\ 5 & -9 & -13 \\ 3 & 1 & -3 \end{vmatrix} = 4 \begin{vmatrix} 1 & -1 & -1 \\ 5 & -9 & -13 \\ 3 & 1 & -3 \end{vmatrix} = 4 \begin{vmatrix} 1 & 0 & -1 \\ 5 & -4 & -13 \\ 3 & 4 & -3 \end{vmatrix} = 4 \begin{vmatrix} 1 & 0 & 0 \\ 5 & -4 & -8 \\ 3 & 4 & 0 \end{vmatrix}$

$= 4 \times 1 \times \begin{vmatrix} -4 & -8 \\ 4 & 0 \end{vmatrix} = 4[(-4 \times 0) - (-8 \times 4)] = 128$

Mathematics beyond the Calculus

Factor a determinant. (Hint – subtract row 3 from rows 1 and 2.)

$$(9)\ \begin{vmatrix} 1 & a & a^2 \\ 1 & b & b^2 \\ 1 & c & c^2 \end{vmatrix} = \begin{vmatrix} 0 & a-c & a^2-c^2 \\ 0 & b-c & b^2-c^2 \\ 1 & c & c^2 \end{vmatrix} = (a-c)(b-c)\begin{vmatrix} 0 & 1 & a+c \\ 0 & 1 & b+c \\ 1 & c & c^2 \end{vmatrix}$$

$$= (a-c)(b-c)(1)\begin{vmatrix} 1 & a+c \\ 1 & b+c \end{vmatrix} = (a-c)(b-c)[(b+c)-(a+c)] = (a-c)(b-c)(b-a)$$

1.4 Minors and Cofactors

Minor ij (M_{ij}) of determinant A is created by striking row i and column j. The *cofactor*, $cf_{ij}(A)$ equals $M_{ij} \times (-1)^{i+j}$.

$$(10)\ A = \begin{vmatrix} a_1 & b_1 & c_1 \\ a_2 & b_2 & c_2 \\ a_3 & b_3 & c_3 \end{vmatrix} \rightarrow M_{12} = \begin{vmatrix} a_2 & c_2 \\ a_3 & c_3 \end{vmatrix} \rightarrow M_{33} = \begin{vmatrix} a_1 & b_1 \\ a_2 & b_2 \end{vmatrix}$$

$$cf_{12}(A) = (-1)^{1+2} M_{12} = -M_{12} \qquad cf_{33}(A) = (-1)^{3+3} M_{33} = M_{33}$$

1.5 Expansion of Determinants

(11) *Laplace Expansions of $n \times n$ determinant A*

$$\text{by row } i \quad \det A = \sum_{j=1}^{n} a_{ij} cf_{ij}(A) \qquad \text{by column } j \quad \det A = \sum_{i=1}^{n} a_{ij} cf_{ij}(A)$$

(12a) expand by row $i \rightarrow \det A = a_{i1} A_{i1} + a_{i2} A_{i2} + \ldots + a_{in} A_{in}$

(12b) expand by column $j \rightarrow \det A = a_{1j} A_{1j} + a_{2j} A_{2j} + \ldots + a_{nj} A_{nj}$

In an expansion of A if the sum of row number plus column number is odd the sign of a term is negative. If the sum of row number plus column number is even the sign of a term is positive. An $r \times c$ determinant can be expanded by rows or by columns. Here are expansions by row 2 and column 2.

$$(13)\ \text{row } 2 \rightarrow A = \begin{vmatrix} a_{11} & a_{12} & a_{13} \\ a_{21} & a_{22} & a_{23} \\ a_{31} & a_{32} & a_{33} \end{vmatrix} = -a_{21}\begin{vmatrix} a_{12} & a_{13} \\ a_{32} & a_{33} \end{vmatrix} + a_{22}\begin{vmatrix} a_{11} & a_{13} \\ a_{31} & a_{33} \end{vmatrix} - a_{23}\begin{vmatrix} a_{11} & a_{12} \\ a_{31} & a_{32} \end{vmatrix}$$

$$= -a_{21} cf_{21}(A) + a_{22} cf_{22}(A) - a_{23} cf_{23}(A)$$

$$(14)\ \text{col } 2 \rightarrow A = \begin{vmatrix} a_{11} & a_{12} & a_{13} \\ a_{21} & a_{22} & a_{23} \\ a_{31} & a_{32} & a_{33} \end{vmatrix} = -a_{12}\begin{vmatrix} a_{21} & a_{23} \\ a_{31} & a_{33} \end{vmatrix} + a_{22}\begin{vmatrix} a_{11} & a_{13} \\ a_{31} & a_{33} \end{vmatrix} - a_{32}\begin{vmatrix} a_{11} & a_{13} \\ a_{21} & a_{23} \end{vmatrix}$$

$$= -a_{12} cf_{12}(A) + a_{22} cf_{22}(A) - a_{32} cf_{32}(A)$$

1 Determinants

Cramer's Rule

The first subscript is the row number. The second subscript is the column number. Determinants are expanded by rows or columns.

Cramer's solutions are expansions by columns where forcing functions replace the column's elements. Note: incorporate minus signs into the a_{ij}'s.

Cramer found responses y_1, y_2 to forcing functions x_1, x_2.

$$\text{if} \quad x_1 = a_{11} y_1 + a_{12} y_2$$
$$\text{and} \quad x_2 = a_{21} y_1 + a_{22} y_2$$
$$\text{Then} \quad \Delta = a_{11} a_{22} - a_{21} a_{12} \quad \text{and}$$

$$y_1 = \frac{\begin{vmatrix} x_1 & a_{12} \\ x_2 & a_{22} \end{vmatrix}}{\begin{vmatrix} a_{11} & a_{12} \\ a_{21} & a_{22} \end{vmatrix}} = \frac{x_1 a_{22} - x_2 a_{12}}{\Delta} \qquad y_2 = \frac{\begin{vmatrix} a_{11} & x_1 \\ a_{21} & x_2 \end{vmatrix}}{\begin{vmatrix} a_{11} & a_{12} \\ a_{21} & a_{22} \end{vmatrix}} = \frac{-x_1 a_{21} + x_2 a_{11}}{\Delta}$$

And, for three responses y_1, y_2, y_3 to forcing functions x_1, x_2, x_3.

$$\text{If } x_1 = a_{11} y_1 + a_{12} y_2 + a_{13} y_3$$
$$x_2 = a_{21} y_1 + a_{22} y_2 + a_{23} y_3$$
$$x_3 = a_{31} y_1 + a_{32} y_2 + a_{33} y_3$$

$$\text{Then } \Delta = a_{11} \Delta_{11} - a_{21} \Delta_{21} + a_{31} \Delta_{31} \quad \textit{(expansion by column 1)}$$
$$\Delta = a_{11}(a_{22} a_{33} - a_{23} a_{32}) - a_{21}(a_{12} a_{33} - a_{13} a_{32}) + a_{31}(a_{12} a_{23} - a_{13} a_{22})$$

$$y_1 = \frac{x_1 \Delta_{11} - x_2 \Delta_{21} + x_3 \Delta_{31}}{\Delta} \quad \textit{(expansion down column 1, rows 1, 2, 3)}$$

$$y_2 = \frac{x_1 \Delta_{12} - x_2 \Delta_{22} + x_3 \Delta_{32}}{\Delta} \quad \textit{(expansion down column 2, rows 1, 2, 3)}$$

$$y_3 = \frac{x_1 \Delta_{13} - x_2 \Delta_{23} + x_3 \Delta_{33}}{\Delta} \quad \textit{(expansion down column 3, rows 1, 2, 3)}$$

Mathematics beyond the Calculus

Problems Factor the determinants

(101) $\begin{vmatrix} 1 & a & a^3 \\ 1 & b & b^3 \\ 1 & c & c^3 \end{vmatrix}$ (102) $\begin{vmatrix} x^2 & x & (y+z) \\ y^2 & y & (x+z) \\ z^2 & z & (x+y) \end{vmatrix}$ (103) $\begin{vmatrix} 1 & 1 & 1 \\ yz & xz & xy \\ x^3 & y^3 & z^3 \end{vmatrix}$

Problems Evaluate the determinants

(104) $\begin{vmatrix} \dfrac{a^2+b^2}{c} & c & c \\ a & \dfrac{b^2+c^2}{a} & a \\ b & b & \dfrac{a^2+c^2}{b} \end{vmatrix}$ (105) $\begin{vmatrix} 1 & b & b & b \\ 1 & a & b & b \\ 1 & b & a & b \\ 1 & b & b & a \end{vmatrix}$

(106) $\begin{vmatrix} 5 & 3 \\ -8 & 12 \end{vmatrix}$ (107) $\begin{vmatrix} 2 & 3 & 4 \\ 4 & 3 & 1 \\ 1 & 2 & 4 \end{vmatrix}$ (108) $\begin{vmatrix} 2 & 1 & -1 & 2 \\ 1 & 3 & 2 & -3 \\ -1 & 2 & 1 & -1 \\ 2 & -3 & -1 & 4 \end{vmatrix}$

(109) $\begin{vmatrix} 5 & -7 \\ 2 & 3 \end{vmatrix}$ (110) $\begin{vmatrix} 3 & -1 & 2 \\ 0 & 5 & 3 \\ 0 & 0 & 1 \end{vmatrix}$ (111) $\begin{vmatrix} 1 & 2 & 3 & 4 \\ -1 & 1 & 2 & 3 \\ 1 & -1 & 1 & 2 \\ -1 & 1 & -1 & 1 \end{vmatrix}$

(112) $det A = \begin{vmatrix} -\lambda & 1 & 1 \\ 1 & -\lambda & 1 \\ 1 & 1 & -\lambda \end{vmatrix}$ (113) $det A = \begin{vmatrix} -\lambda & 0 & 1 \\ 1 & -\lambda & -3 \\ 0 & 1 & 3-\lambda \end{vmatrix}$

Problem Solve for x

(114) $\begin{vmatrix} 2 & -1 & -1 \\ -x & 1 & -3 \\ -2 & x & 1 \end{vmatrix} = 0$

Problem Use Cramer's rule, solve

(115) $\begin{bmatrix} 1 & 2 & -1 \\ -1 & 1 & 2 \\ 2 & -1 & 1 \end{bmatrix} \begin{bmatrix} x \\ y \\ z \end{bmatrix} = \begin{bmatrix} 3 \\ 2 \\ 1 \end{bmatrix}$

2 Finite Matrices

Matrix notation is a means of writing linear equations in an orderly form. The advantages are significant when several linear systems have to be combined and solved.

2.1 The (7, 4) Decoder[1]

> The Decoder shows how matrices are used in an application.

A transmitted code word is decoded when received. Uncommon common sense says *Parity equations were used to encode so use them to decode*.

Received code word At the receiver the available information is the received code word R. **If** R could be compared to the transmitted code word C, **then** the bit by bit sum modulo 2 of R and C is the error word E.

In other words, the received word R is the sum[2] modulo 2, without carries, of transmitted code word C and error word E. The following error word E reveals that there are 2 errors in R. However E is not known in practice.

Transmitted block C	1001100	*Transmitted block C*	1001100
Received block R	1000101	*Error block E*	+0001001 $R = C + E$
Bits in error E	0001001	*Received block R*	1000101

The H Matrix and Syndrome Matrix S The next step in the process implementing a code is the design of the decoding matrix. The decoding matrix is referred to as the parity matrix H, which Hamming produced for us in Section 1.2. Pretending we do not know that, we will construct H directly from the parity equations. First define a matrix operation.

$$(17a) \quad B = \begin{bmatrix} 1 & 2 \\ 3 & 4 \\ 5 & 6 \end{bmatrix} \text{ is the transpose of } A = \begin{bmatrix} 1 & 3 & 5 \\ 2 & 4 & 6 \end{bmatrix} \quad (17b) \quad B = A^T$$

---

[1] Nicholas L. Pappas "Error correction code Design" ISBN 978 1511 813 860
[2] Modulo 2 arithmetic – addition 0+0=0, 0+1=1, 1+0=1, 1+1=0 and multiplication 0×0=0, 0×1=0, 1×0=0, 1×1=1

Mathematics beyond the Calculus

(18) $\quad R = C + E$

(19) $\quad S = HR^T = HC^T + HE^T = 0 + HE^T = HE^T$

where S is referred to as the syndrome

Observe that the syndrome S only depends on the error word E and not on the transmitted code word C. $HC^T = 0$ because there are no errors in code word C. Since r=3, there are three parity equations to calculate and so matrix S has 3 terms in the form of a 3×1 matrix. H uses the parity equations so that $HR^T=0$ when a received code word R has no errors.

Three parity equations imply three rows for H. Matrix multiplication of H times C^T *scans* rows of H calculating the three parity equations by summing terms. No errors require the three sums to equal 0, because of the definition of the parity equations.

(20) $\quad \begin{aligned} 0 &= k_3 + k_2 + k_1 + r_2 \\ 0 &= k_3 + k_2 + k_0 + r_1 \\ 0 &= k_3 + k_1 + k_0 + r_0 \end{aligned} \Rightarrow HC^T = \begin{aligned} c_7 + c_6 + c_5 + c_4 + 0 + 0 + 0 &= 0 \\ c_7 + c_6 + 0 + 0 + c_3 + c_2 + 0 &= 0 \\ c_7 + 0 + c_5 + 0 + c_3 + 0 + c_1 &= 0 \end{aligned}$

(21) $\quad H = \begin{bmatrix} 1111000 \\ 1100110 \\ 1010101 \end{bmatrix}$

Consequently the syndrome equations are

(22) $\quad S = H \times R^T = \begin{bmatrix} s_2 \\ s_1 \\ s_0 \end{bmatrix} = \begin{bmatrix} 1111000 \\ 1100110 \\ 1010101 \end{bmatrix} \times \begin{bmatrix} b_7 & b_6 & b_5 & b_4 & b_3 & b_2 & b_1 \end{bmatrix}^T$

Note that the syndrome S is a 3 bit number $s_2 s_1 s_0$, which can have 8 different values ($2^r = 2^3 = 8$). However there are $2^7=128$ different received words R. Many R contain 2 or more errors that this code cannot process. These are referred to as decoding failures, which are *not* detected.

Error Detection Syndrome equations detect the *presence* of errors.
Error correction The syndrome information allows for correction of errors in the received word R. In more complex codes, such as BCH, the syndrome information is processed in a complex way to identify and correct the bits in error.

Important The syndrome in the Hamming code process directly produces the bit error position number. The price paid is that the Hamming code presented here fails when there is more than 1 error. The need to correct more than one error was resolved by the BCH codes.

2 Finite Matrices

Three examples of syndrome calculations with 0, 1, and 2 errors show that the *hardware interpretations* are as follows.

(23) *Example* : $R = 1001100 + 0000000$ (*no errors*)

$$S = H \times R^T = \begin{bmatrix} 1111000 \\ 1100110 \\ 1010101 \end{bmatrix} \times \begin{bmatrix} 1 \\ 0 \\ 0 \\ 1 \\ 1 \\ 0 \\ 0 \end{bmatrix} = \begin{bmatrix} 0 \\ 0 \\ 0 \end{bmatrix}$$

interpret S as 000, *i.e. no errors*

(24) *Example* : $R = 1001100 + 0000001$ (*one error*)

$$S = H \times R^T = \begin{bmatrix} 1111000 \\ 1100110 \\ 1010101 \end{bmatrix} \times \begin{bmatrix} 1 \\ 0 \\ 0 \\ 1 \\ 1 \\ 0 \\ 1 \end{bmatrix} = \begin{bmatrix} 0 \\ 0 \\ 1 \end{bmatrix}$$

interpret S as 001, *i.e. error in bit position* 1

(25) *Example*: $R = 1001100 + 0100001 = 1101101$ (*two errors, bits 6 and 1*)

$$S = H \times R^T = \begin{bmatrix} 1111000 \\ 1100110 \\ 1010101 \end{bmatrix} \times \begin{bmatrix} 1 \\ 1 \\ 0 \\ 1 \\ 1 \\ 0 \\ 1 \end{bmatrix} = \begin{bmatrix} 1 \\ 1 \\ 1 \end{bmatrix}$$

interpret S as 111, *i.e. error in bit position 7, which is not correct*

Note The syndrome report *bit seven is the error bit* is not correct. Bit 7 is inverted by the error correction hardware and the received word R=1101101 becomes 0101101. This is a decoding failure that *cannot* be detected by calculating the syndrome S of the corrected word 0101101, because S still equals zero! Try it.

Mathematics beyond the Calculus

2.2 Definition

An array of rc numbers arranged as r rows and c columns is referred to as a matrix. Thus

(1) $M_{2 \times 2} = \begin{bmatrix} m_{11} & m_{12} \\ m_{21} & m_{22} \end{bmatrix}$ $\quad M_{r \times c} = \begin{bmatrix} m_{11} & m_{12} & m_{13} & \cdots & m_{1c} \\ m_{21} & m_{22} & m_{23} & \cdots & m_{2c} \\ \vdots & \vdots & \vdots & \vdots & \vdots \\ m_{r1} & m_{r2} & m_{r3} & \cdots & m_{rc} \end{bmatrix}$

are matrices. The m_{ij} are elements of the matrix M. The element in matrix M row i and column j is m_{ij}, and this is indicated by $M(m_{ij})$. The element m_{ij} is said to have row index i, and column index j. Note that the row index i is always listed first.

For example

(2a) $A = \begin{bmatrix} 3 & 8 & 3 & 0 \\ 9 & 5 & 2 & 13 \\ 7 & 6 & 5 & 10 \end{bmatrix}$ (2b) $B = \begin{bmatrix} 2 & 0 \\ -9 & 5 \end{bmatrix}$ (2c) $C = \begin{bmatrix} 6 & 2 \\ -1 & 4 \\ 5 & 3 \end{bmatrix}$

Where A has 3 × 4, B has 2 × 2, and C has 3 × 2 rows and columns. In the above examples $a_{23} = 2$, $b_{12} = 0$, and $c_{31} = 5$.

Definition – An r × c matrix is an array of rc elements consisting of r rows and c columns. The matrix is referred to as a rectangular matrix when $r \neq c$, and a square matrix when $r = c$.

Definition – Two matrices are equal if, and only if, all corresponding elements of them are equal.

2 Finite Matrices

2.3 Matrix Multiplication

The equation y = bx may be interpreted as multiplication by b *transforms* x into y.

Matrices arise in the study of linear systems. For example consider two linear equations where r = 2 and c = 3.
(3a) $w_1 = a_{11}x_1 + a_{12}x_2 + a_{13}x_3$
(3b) $w_2 = a_{21}x_1 + a_{22}x_2 + a_{23}x_3$

These equations may be interpreted as defining a transformation. Multiplication by coefficients a_{ij} *transforms* the x_j into the w_i.

Casting the equations into matrix form the associated matrices are

(4) $A = \begin{bmatrix} a_{11} & a_{12} & a_{13} \\ a_{21} & a_{22} & a_{23} \end{bmatrix} \quad X = \begin{bmatrix} x_1 \\ x_2 \\ x_3 \end{bmatrix} \quad W = \begin{bmatrix} w_1 \\ w_2 \end{bmatrix} \quad \text{and} \quad W = AX$

The following definition of a matrix product, while apparently arbitrary, produces the correct result when the number of columns of R equal the number of rows of C. Then R and C are said to be *conformable*.

If

$(5a) \; R = row_i(M) = \begin{bmatrix} m_{i1} & m_{i2} & m_{i3} & \ldots & m_{ic} \end{bmatrix} \quad (5b) \; C = col_j(M) = \begin{bmatrix} m_{1j} \\ m_{2j} \\ m_{3j} \\ \vdots \\ m_{cj} \end{bmatrix}$

then

$(6) \; RC = \begin{bmatrix} m_{i1} & m_{i2} & m_{i3} & \ldots & m_{ic} \end{bmatrix} \times \begin{bmatrix} m_{1j} \\ m_{2j} \\ m_{3j} \\ \vdots \\ m_{cj} \end{bmatrix}$

$RC = m_{i1}m_{1j} + m_{i2}m_{2j} + \ldots + m_{ic}m_{cj} = \sum_{x=1}^{c} m_{ix}m_{xj}$

Mathematics beyond the Calculus

To see a multiplication example transform Y into X of equation 4.

(7) $x_1 = b_{11}y_1 + b_{12}y_2$
$x_2 = b_{21}y_1 + b_{22}y_2$
$x_3 = b_{31}y_1 + b_{32}y_2$

(8) $X = BY \quad X = \begin{bmatrix} x_1 \\ x_2 \\ x_3 \end{bmatrix} \quad B = \begin{bmatrix} b_{11} & b_{12} \\ b_{21} & b_{22} \\ b_{31} & b_{32} \end{bmatrix} \quad Y = \begin{bmatrix} y_1 \\ y_2 \end{bmatrix}$ and $W = AX = ABY$

(9) let $C = AB = \begin{bmatrix} row_1(A) \times col_1(B) & row_1(A) \times col_2(B) \\ row_2(A) \times col_1(B) & row_2(A) \times col_2(B) \end{bmatrix}$ and $W = CY$

where

(10) $C = AB = \begin{bmatrix} [a_{11} \; a_{12} \; a_{13}] \times \begin{bmatrix} b_{11} \\ b_{21} \\ b_{31} \end{bmatrix} & [a_{11} \; a_{12} \; a_{13}] \times \begin{bmatrix} b_{12} \\ b_{22} \\ b_{32} \end{bmatrix} \\ [a_{21} \; a_{22} \; a_{23}] \times \begin{bmatrix} b_{11} \\ b_{21} \\ b_{31} \end{bmatrix} & [a_{21} \; a_{22} \; a_{23}] \times \begin{bmatrix} b_{12} \\ b_{22} \\ b_{32} \end{bmatrix} \end{bmatrix}$

Here is a numerical example

(11) $A = \begin{bmatrix} 3 & 0 & -1 \\ 6 & -2 & 4 \end{bmatrix} \quad B = \begin{bmatrix} 5 & -1 \\ -3 & 0 \\ 2 & 7 \end{bmatrix}$

(12) $C_{2\times 2} = A_{2\times 3} B_{3\times 2} = \begin{bmatrix} row_1(A) \times col_1(B) & row_1(A) \times col_2(B) \\ row_2(A) \times col_1(B) & row_2(A) \times col_2(B) \end{bmatrix}$

(13) $C = AB = \begin{bmatrix} [3 \; 0 \; -1] \times \begin{bmatrix} 5 \\ -3 \\ 2 \end{bmatrix} & [3 \; 0 \; -1] \times \begin{bmatrix} -1 \\ 0 \\ 7 \end{bmatrix} \\ [6 \; -2 \; 4] \times \begin{bmatrix} 5 \\ -3 \\ 2 \end{bmatrix} & [6 \; -2 \; 4] \times \begin{bmatrix} -1 \\ 0 \\ 7 \end{bmatrix} \end{bmatrix} = \begin{bmatrix} 13 & -10 \\ 44 & 22 \end{bmatrix}$

Emphasis – The matrix product AB requires that the matrices are *conformable*: the number of columns of A must equal the number of rows of B. A must be m × c and B must be c × n. Then C = AB is m × n. The matrix product BA is not *conformable*. This why it has no meaning.

2.4 Matrix Sums and Differences

Given two linear transformations of variables x into variables y, the sum and difference of the two transformations can be formed. If

(14a) $y_1 = 5x_1 + 7x_2 - x_3$ (15a) $w_1 = x_1 - 2x_2 + 6x_3$
(14b) $y_2 = 3x_1 + 9x_2 + 2x_3$ (15b) $w_2 = 7x_1 + 2x_2 + x_3$

then

(16a) $z_1 = y_1 + w_1 = 6x_1 + 5x_2 + 5x_3$
(16b) $z_2 = y_2 + w_2 = 10x_1 + 11x_2 + 3x_3$

I.e. if $A = (a_{ij})$ and $B = (b_{ij})$ are both $r \times c$ matrices, then $A \pm B$ is defined to be the $r \times c$ matrix $S = (s_{ij})$ where $s_{ij} = a_{ij} \pm b_{ij}$.

Emphasis – both matrices must be of the *same* size.

Example:

$$(17)\ S = A + B = \begin{bmatrix} 3 & 8 & 3 & 0 \\ 9 & 5 & 2 & 13 \\ 7 & 6 & 5 & 10 \end{bmatrix} + \begin{bmatrix} 3 & 1 & 0 & 5 \\ 8 & 6 & 2 & 4 \\ 0 & 9 & 5 & 10 \end{bmatrix} = \begin{bmatrix} 6 & 9 & 3 & 5 \\ 17 & 11 & 4 & 17 \\ 7 & 15 & 10 & 20 \end{bmatrix}$$

2.5 The Adjoint Matrix

Replace each element of a matrix B by its cofactor. Then transpose this matrix, which is referred to as the adjoint matrix of B, adj(B). For example

$$(18)\ B = \begin{bmatrix} 2 & 4 & 6 \\ -1 & 2 & 3 \\ 1 & 4 & 9 \end{bmatrix} \to cf_{11} = \begin{vmatrix} 2 & 3 \\ 4 & 9 \end{vmatrix} = 6 \to cf_{12} = -\begin{vmatrix} -1 & 3 \\ 1 & 9 \end{vmatrix} = 12$$

$$\to cf_{13} = \begin{vmatrix} -1 & 2 \\ 1 & 4 \end{vmatrix} = -6 \to cf_{21} = -\begin{vmatrix} 4 & 6 \\ 4 & 9 \end{vmatrix} = -12 \to cf_{22} = \begin{vmatrix} 2 & 6 \\ 1 & 9 \end{vmatrix} = 12$$

$$\to cf_{23} = -\begin{vmatrix} 2 & 4 \\ 1 & 4 \end{vmatrix} = -4 \to cf_{31} = \begin{vmatrix} 4 & 6 \\ 2 & 3 \end{vmatrix} = 0 \to cf_{32} = -\begin{vmatrix} 2 & 6 \\ -1 & 3 \end{vmatrix} = -12$$

$$\to cf_{33} = \begin{vmatrix} 2 & 4 \\ -1 & 2 \end{vmatrix} = 8$$

Then the adjoint of B is as follows.

$$(19)\ adjB = \begin{bmatrix} cf_{11} & cf_{12} & cf_{13} \\ cf_{21} & cf_{22} & cf_{23} \\ cf_{31} & cf_{32} & cf_{33} \end{bmatrix}^T = \begin{bmatrix} cf_{11} & cf_{21} & cf_{31} \\ cf_{12} & cf_{22} & cf_{32} \\ cf_{13} & cf_{23} & cf_{33} \end{bmatrix} = \begin{bmatrix} 6 & -12 & 0 \\ 12 & 12 & -12 \\ -6 & -4 & 8 \end{bmatrix}$$

Mathematics beyond the Calculus

2.6 Inverse of a Square Matrix (r = c)

Only square matrices have inverses. The inverse is related to the adjoint as follows.

(20) $inverseB = \dfrac{1}{detB} adjB = \dfrac{1}{24}\begin{bmatrix} 6 & -12 & 0 \\ -12 & 12 & -12 \\ -6 & -4 & 8 \end{bmatrix} = \begin{bmatrix} \frac{1}{4} & -\frac{1}{2} & 0 \\ -\frac{1}{2} & \frac{1}{2} & -\frac{1}{2} \\ -\frac{1}{4} & -\frac{1}{6} & \frac{1}{3} \end{bmatrix} = B^{-1}$

The system of equations y can be written as the matrix equation Y=AX.
(21a) $y_1 = a_{11}x_1 + a_{12}x_2 + a_{13}x_3$
(21b) $y_2 = a_{21}x_1 + a_{22}x_2 + a_{23}x_3$
(21c) $y_3 = a_{31}x_1 + a_{32}x_2 + a_{33}x_3$

(22) $Y = AX \;\rightarrow\; \begin{bmatrix} y_1 \\ y_2 \\ y_3 \end{bmatrix} = \begin{bmatrix} a_{11} & a_{12} & a_{13} \\ a_{21} & a_{22} & a_{23} \\ a_{31} & a_{32} & a_{33} \end{bmatrix} \times \begin{bmatrix} x_1 \\ x_2 \\ x_3 \end{bmatrix}$

If the number $\Delta = detA \neq 0$ is the determinant of the system and the A_{ij} (another way the cf_{ij} are written) are the cofactors of Δ, then the system solution is

(23a) $x_1 = \dfrac{A_{11}}{\Delta}y_1 + \dfrac{A_{21}}{\Delta}y_2 + \dfrac{A_{31}}{\Delta}y_3$

(23b) $x_2 = \dfrac{A_{12}}{\Delta}y_1 + \dfrac{A_{22}}{\Delta}y_2 + \dfrac{A_{32}}{\Delta}y_3$

(23c) $x_3 = \dfrac{A_{13}}{\Delta}y_1 + \dfrac{A_{23}}{\Delta}y_2 + \dfrac{A_{33}}{\Delta}y_3$

where Aij is $(-1)^{i+j}$ times the minor of the determinant (strike row i, column j)

(24) $A^{-1} = \begin{bmatrix} \frac{A_{11}}{\Delta} & \frac{A_{21}}{\Delta} & \frac{A_{31}}{\Delta} \\ \frac{A_{12}}{\Delta} & \frac{A_{22}}{\Delta} & \frac{A_{32}}{\Delta} \\ \frac{A_{13}}{\Delta} & \frac{A_{23}}{\Delta} & \frac{A_{33}}{\Delta} \end{bmatrix} = \dfrac{1}{\Delta}\begin{bmatrix} A_{11} & A_{21} & A_{31} \\ A_{12} & A_{22} & A_{32} \\ A_{13} & A_{23} & A_{33} \end{bmatrix}$

and A^{-1} is the inverse of A $\;\rightarrow\; A^{-1}A = AA^{-1} = \begin{bmatrix} 1 & 0 & 0 \\ 0 & 1 & 0 \\ 0 & 0 & 1 \end{bmatrix} = I$

Emphasis – In matrix algebra division does not exist. Instead one multiplies by the inverse.
(25) $Y = AX \;\rightarrow\; A^{-1}Y = A^{-1}AX = IX = X \;\rightarrow\; X = A^{-1}Y$

2.7 Inverse Reversal Rule

If A and B are non singular n × n matrices, then AB is also non singular and $(AB)^{-1} = (B^{-1}A^{-1})$. Here is why.

(26a) $(AB)(B^{-1}A^{-1}) = A(BB^{-1})A^{-1} = AIA^{-1} = AA^{-1} = I$

(26b) $(B^{-1}A^{-1})(AB) = B^{-1}(A^{-1}A)B = B^{-1}IB = B^{-1}B = I$

(26c) *therefore* $(AB)(AB)^{-1} = (AB)(B^{-1}A^{-1})$ → $(AB)^{-1} = (B^{-1}A^{-1})$

(26c) *and* $(AB)^{-1}(AB) = (B^{-1}A^{-1})(AB)$ → *again* $(AB)^{-1} = (B^{-1}A^{-1})$

2.8 Cancellation of Common Factors

Two facts – Generally the product of two matrices does not commute. A *singular* matrix whose determinant Δ equals 0 does *not* have an inverse. Consequently one has to be careful before canceling common factors.

(27) *If $AC = BC$ and C has an inverse (it is non singular), then*
$ACC^{-1} = BCC^{-1}$ → $AI = BI$ → $A = B$

(28) *If $AC = CB$ and C has an inverse (it is non singular), then*
$ACC^{-1} = CBC^{-1}$ → $AI = CBC^{-1}$ → $A = CBC^{-1}$ → $A \neq B$

2.9 Complex elements

When elements of A are complex the matrix is referred to as a complex matrix. For example

(29) *if* $A = \begin{bmatrix} i3 & 4-i \\ 9+i2 & 5 \\ 7 & -i6 \end{bmatrix}$ *then conjugate* $A = \bar{A} = \begin{bmatrix} -i3 & 4+i \\ 9-i2 & 5 \\ 7 & i6 \end{bmatrix}$

2.10 Transpose of a Matrix

The transpose of r × c matrix A is written as A^T. A^T is formed by converting rows of A into columns of A^T. Elements may be real and, or complex. For example

(30) $\quad if\ A = \begin{bmatrix} 3 & 8 & 3 & 0 \\ 9 & 5 & 2 & 13 \\ 7 & 6 & 5 & 10 \end{bmatrix} \quad then\ A^T = \begin{bmatrix} 3 & 9 & 7 \\ 8 & 5 & 6 \\ 3 & 2 & 5 \\ 0 & 13 & 10 \end{bmatrix}$

The basic properties of transposes are as follows.
(31) $\quad (A^T)^T = A \quad (AB)^T = B^T A^T \quad (A+B)^T = A^T + B^T \quad (A^T)^{-1} = (A^{-1})^T$

2.11 Scalar Multiplication

The basic properties of scalars are as follows.
(32) $\quad 1A = A \quad 0A = 0$
$(x+y)A = xA + yA \quad x(A+B) = xA + xB$
$x(yA) = (xy)A \quad A(yB) = y(AB)$

2.12 Linear Dependence

Consider matrix A with 3 rows.

(18) $\quad Y = AX \quad \rightarrow \quad \begin{bmatrix} y_1 \\ y_2 \\ y_3 \end{bmatrix} = \begin{bmatrix} a_{11} & a_{12} & a_{13} \\ a_{21} & a_{22} & a_{23} \\ a_{31} & a_{32} & a_{33} \end{bmatrix} \times \begin{bmatrix} x_1 \\ x_2 \\ x_3 \end{bmatrix}$

If the 3 rows are related in some way, then there may be numbers λ_1 and λ_2 for which
(33) $\quad row_3 = \lambda_1 row_1 + \lambda_2 row_2$
When the λ numbers are not zero some of the rows are linearly dependent.

2.13 Rank of a Matrix

The rank of a matrix equals the number of linearly *independent* rows of the matrix. Rank is the most important property. Another way to express the idea of rank is this. If A has the property that t is the largest integer such that A has a $t \times t$ sub-matrix with non zero determinant then t = rank of A.

2 Finite Matrices

2.14 Vandermonde Matrices

Finding and verifying that any matrix has the required rank is a very difficult problem. *The known rank of the very useful Vandermonde matrices solves this problem immediately.*

Rank r tells us the r rows in an r×r matrix are independent. Rank less than r tells us the equations represented by the matrix are inconsistent, and thus do not have a solution. The rank of a matrix is equal to the number of independent rows in the matrix. Emphasis: each row represents a linear equation.

The Vandermonde matrices are one of various families of matrices that have special structures whose rank for any order is readily determined! This is truly amazing and fortunate. A Vandermonde matrix has a special form *whose rank is known. The terms of the rows are powers of the terms y_k in the first row.*

(34) $V_{n \times n} = \begin{bmatrix} y_1 & y_2 & \cdots & y_n \\ y_1^2 & y_2^2 & \cdots & y_n^2 \\ \vdots & \vdots & & \vdots \\ y_1^n & y_2^n & \cdots & y_n^n \end{bmatrix}$ (all y_i are distinct and non zero)

Reduced Vandermonde matrix The *reduced* matrix $V_{t \times n}$ has rank t. The reduced matrix consists of the *odd rows* of a Vandermonde matrix, and columns 1 to n. The reduced matrix is not a Vandermonde matrix. The matrix can be reduced, because modulo 2 arithmetic makes even powers of *binary n-tuples y* equal to odd powers of *binary n-tuples y*. This is *not* obvious. A binary n-tuple is a string of n bits. E.g. 101101 is a 6-tuple.

(35) $V_{t \times n} = \begin{bmatrix} y_1^1 & y_2^1 & \cdots & y_n^1 \\ y_1^3 & y_2^3 & \cdots & y_n^3 \\ \vdots & \vdots & & \vdots \\ y_1^{2t-1} & y_2^{2t-1} & \cdots & y_n^{2t-1} \end{bmatrix}$

t	2t-1	odd rows
1	1	1
2	3	1, 3
3	5	1, 3, 5

Furthermore, any set of t columns of a t×n reduced matrix contains a t×t matrix (since n≥t), which is related to the t×t Vandermonde matrix. The magic words *for applications* are:

The t×t matrix within the t×n matrix derived from a Vandermonde n×n matrix has rank t.

Mathematics beyond the Calculus

Problem 201

Compute AB and BA where $A = \begin{bmatrix} 1 & 2 & 3 \\ 4 & 5 & 6 \end{bmatrix}$ $B = \begin{bmatrix} 1 & 0 \\ 2 & 3 \\ 0 & 4 \end{bmatrix}$

Problem 202

If $AC = CA$ and $BC = CB$ where $C = \begin{bmatrix} 0 & 1 \\ -1 & 0 \end{bmatrix}$ then show that $AB = BA$

Problem 203 Write the matrix equation corresponding to these equations.
$12 = 2x_1 + 5x_2 + 7x_3$
$0 = 3x_1 - 6x_2 + x_3$
$5 = -x_1 + 7x_2 + 3x_3$

Problem 204

Compute the inverse of $\begin{bmatrix} d_{11} & 0 & 0 \\ 0 & d_{22} & 0 \\ 0 & 0 & d_{33} \end{bmatrix}$

Problem 205
compute A^{-1} when $A^3 - 4A^2 + A + 3I = 0$

Problem 206
solve for A when $(A^T B)^{-1} - (B^T A^{-1})^{-1} + (B^{-1} A^T)^T = I$

Problem 207
if A^T and B^T commute show that A and B commute

Problem 208
Reference equations 18, 19, 20. Show that $BB^{-1} = I$ and that $B^{-1}B = I$

3 Eigenvalues and Eigenvectors

Suppose that A is a square n × n matrix. Then a nonzero vector x is an eigenvector and a number λ is its eigenvalue when $Ax = \lambda x$. In matrix format the equation is

(1) $A_{n\times n} x_{n\times 1} = \lambda x_{n\times 1} \rightarrow Ax = \lambda x$

Geometrically this means that x is in the same direction as Ax, because multiplying a vector by a real number changes its length, but not its direction.

In other words an eigenvector x of matrix operator A is a vector, which is not altered in direction by the operator. The vector x may be altered in magnitude. It is a solution of the equation $Ax = \lambda x$.

Find Eigenvalues for 2 × 2 and 3 × 3 matrices.

If $x \neq 0$ then equation (1) can be rewritten as

(2) $Ax - \lambda x = 0 \rightarrow (A - \lambda I)x = 0 \rightarrow \text{determinant} |A - \lambda I| = 0$

If A is 2×2 then we can find eigenvalues and eigenvectors as follows.

(3a) Let $A = \begin{bmatrix} 1 & 4 \\ 3 & 5 \end{bmatrix}$ $A - \lambda I = \begin{bmatrix} 1 & 4 \\ 3 & 5 \end{bmatrix} - \lambda \begin{bmatrix} 1 & 0 \\ 0 & 1 \end{bmatrix} = \begin{bmatrix} 1 & 4 \\ 3 & 5 \end{bmatrix} - \begin{bmatrix} \lambda & 0 \\ 0 & \lambda \end{bmatrix} = \begin{bmatrix} 1-\lambda & 4 \\ 3 & 5-\lambda \end{bmatrix}$

(3b) then $\det(A - \lambda I) = (1-\lambda)(5-\lambda) - (4\times 3) = \lambda^2 - 6\lambda - 7 = (\lambda+1)(\lambda-7)$

This is referred to as the characteristic equation. The eigenvalues of the matrix A are 7 and –1.

For any eigenvalue λ, a corresponding eigenvector is a vector x such that

(4a) $Ax = \lambda x \rightarrow \begin{bmatrix} 1 & 4 \\ 3 & 5 \end{bmatrix} \begin{bmatrix} x_1 \\ x_2 \end{bmatrix} = \lambda \begin{bmatrix} x_1 \\ x_2 \end{bmatrix} \rightarrow \begin{bmatrix} x_1 + 4x_2 \\ 3x_1 + 5x_2 \end{bmatrix} - \begin{bmatrix} \lambda x_1 \\ \lambda x_2 \end{bmatrix} = 0$

(4b) $\begin{bmatrix} (1-\lambda)x_1 + 4x_2 \\ 3x_1 + (5-\lambda)x_2 \end{bmatrix} = 0$

(4c) for $\lambda = 7$ $Ax - 7x = \begin{bmatrix} -6x_1 + 4x_2 \\ 3x_1 - 2x_2 \end{bmatrix} = 0 \rightarrow x = \begin{bmatrix} x_1 \\ x_2 \end{bmatrix} = \begin{bmatrix} 2 \\ 3 \end{bmatrix}$

(4d) for $\lambda = -1$ $Ax + x = \begin{bmatrix} 2x_1 + 4x_2 \\ 3x_1 + 6x_2 \end{bmatrix} = 0 \rightarrow x = \begin{bmatrix} x_1 \\ x_2 \end{bmatrix} = \begin{bmatrix} -2 \\ 1 \end{bmatrix}$

Mathematics beyond the Calculus

Powers of A in $Ax=\lambda x$

(5) $Ax = \lambda x \rightarrow AAx = A\lambda x = \lambda Ax = \lambda\lambda x \rightarrow A^2 x = \lambda^2 x \rightarrow A^n x = \lambda^n x$

For example find the eigenvalues and eigenvectors of $A = \begin{bmatrix} 1 & -1 \\ 1 & 1 \end{bmatrix}$

The characteristic polynomial is

(6) $\det(A - \lambda I) = \begin{bmatrix} 1-\lambda & -1 \\ 1 & 1-\lambda \end{bmatrix} = (1-\lambda)^2 + 1 = \lambda^2 - 2\lambda + 2$

(7) $\lambda^2 - 2\lambda + 2 = 0 \rightarrow \lambda_1, \lambda_2 = \dfrac{-2 \pm \sqrt{4-8}}{2} = \dfrac{-2 \pm 2\sqrt{-1}}{2} = 1 \pm i$

For any eigenvalue A, a corresponding eigenvector is a vector x such that

(8a) $Ax = \lambda x \rightarrow \begin{bmatrix} 1 & -1 \\ 1 & 1 \end{bmatrix}\begin{bmatrix} x_1 \\ x_2 \end{bmatrix} = \lambda \begin{bmatrix} x_1 \\ x_2 \end{bmatrix} \rightarrow \begin{bmatrix} x_1 - x_2 \\ x_1 + x_2 \end{bmatrix} - \begin{bmatrix} \lambda x_1 \\ \lambda x_2 \end{bmatrix} = 0$

(8b) $\begin{bmatrix} (1-\lambda)x_1 - x_2 \\ x_1 + (1-\lambda)x_2 \end{bmatrix} = 0$

(8c) for $\lambda = 1+i$ $\quad Ax - (1+i)x = \begin{bmatrix} (1-1-i)x_1 - x_2 \\ x_1 + (1-1-i)x_2 \end{bmatrix} = \begin{bmatrix} -ix_1 - x_2 \\ x_1 - ix_2 \end{bmatrix} = 0 \rightarrow x = \begin{bmatrix} i \\ 1 \end{bmatrix}$

(8d) for $\lambda = 1-i$ $\quad Ax + -(1-i)x = \begin{bmatrix} (1-1+i)x_1 - x_2 \\ x_1 + (1-1+i)x_2 \end{bmatrix} = \begin{bmatrix} ix_1 - x_2 \\ x_1 + ix_2 \end{bmatrix} = 0 \rightarrow x = \begin{bmatrix} -i \\ 1 \end{bmatrix}$

Computing eigenvalues requires factoring a polynomial, which can be a formidable task. Galois proved about 1818 that it's impossible to express the roots of a general polynomial of degree five or higher using radicals of the coefficients. That is why formulas like the quadratic formula for polynomials of degree 2 do not exist for degree 5 or greater. Almost all practical eigenvalue computations are accomplished by numerical methods.

Find the eigenvalues and eigenvectors
Problem 301
$A = \begin{bmatrix} -3 & 1 & -3 \\ 20 & 3 & 10 \\ 2 & -2 & 4 \end{bmatrix}$

Problem 302
$A = \begin{bmatrix} 3 & 6 & -8 \\ 0 & 0 & 6 \\ 0 & 0 & 2 \end{bmatrix}$

4 The Laplace Transform

Ordinary and partial differential equations are mathematical statements of physical laws involving rates of change of one or more variables with respect to other variables such as how an electric circuit voltage varies with time. The state of the system at time t = 0 and positions such as x = 0 is described by initial conditions related to t and x.

The Laplace Transform is the basis for a straightforward process that solves ordinary and partial differential equations (Chapters 7, 8, 9). The solution starts by transforming the equations into subsidiary equations, which are ordinary differential equations. The subsidiary equations are solved for the variables of interest by a second Laplace transform or by a differential operator process (Chapter 7). Initial conditions are invoked, and the solved subsidiary equations are inverse transformed back to the original domain *as a solution* of the original problem.

4.1 The Laplace Transform

Let f(t) be a real or complex valued function of t for t > 0, and p = σ+jω be a complex variable used as a parameter. Variable t usually represents time, but may represent anything. The Laplace transform of f(t) is defined as

(1) $$F(p) = \mathscr{L}[f(t)] = \int_0^\infty f(t) e^{-pt} dt$$

The symbol $\mathscr{L}$ is the Laplace transformation *operator* on function f(t) to produce a new function F(p) = $\mathscr{L}$[f(t)]. Equations in the *variable t domain* are transformed into equations in the *complex frequency variable p domain*.

Inverse transform If we use the Laplace transform to solve a problem in the *complex frequency P* domain, then we need an inverse transform to return to the t domain.

(2) $$f(t) = \mathscr{L}^{-1}[F(p)] = \frac{1}{2\pi i} \int_{\sigma-j\infty}^{\sigma+j\infty} F(p) e^{tp} dp$$

A return from the *complex frequency p domain* to the t domain is achieved by performing the inverse operation. The operation is known as the Inverse Laplace Transform (equation 2). This integral for calculating the inverse transform is a contour integration in the p plane (Chapter 6).

4.2 Prologue

Here we show how transforms are used in an application: RLC Circuit Transient Response[3]

RLC circuit responses vary according to the circuit damping or Q (there may be more than one Q in complex circuits). The responses take the forms $e^{-\alpha t}+e^{-\beta t}$, $e^{-\alpha t}$, or $e^{-\alpha t}\sin \beta t$. The cases are presented by analyzing a series RLC circuit with one each R, L, and C in series. Assume the initial stored energy is zero.

Figure 705 RLC Circuit

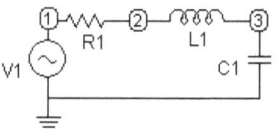

(30) $\quad v_1(t) = v_L + v_R + v_C = L\dfrac{di(t)}{dt} + Ri(t) + \dfrac{1}{C}\displaystyle\int_0^t i(x)dx$

(31) $\quad \mathscr{L}[v_1(t)] = \left[pL + R + \dfrac{1}{pC}\right]\mathscr{L}[i(t)] \quad [v(0) = 0,\ i(0) = 0]$

(32) $\quad \mathscr{L}[i(t)] = \dfrac{1}{pL + R + \dfrac{1}{pC}}\mathscr{L}[v_1(t)] \quad \Rightarrow \quad Z(p) = pL + R + \dfrac{1}{pC}$

(33) $\quad \text{if } v_1(t) = V_m u(t) \text{ then } \mathscr{L}[v_1(t)] = \dfrac{V_m}{p}$

(34) $\quad \mathscr{L}[i(t)] = \dfrac{1}{pL + R + \dfrac{1}{pC}} \cdot \dfrac{V_m}{p} = \dfrac{1}{p^2 + p\dfrac{R}{L} + \dfrac{1}{LC}} \cdot \dfrac{V_m}{L} = \dfrac{V_m}{L}\dfrac{1}{(p+p_1)(p+p_2)}$

(35) $\quad \text{where } p_1, p_2 = -\alpha \mp \beta = -\dfrac{R}{2L}\left[1 \pm \sqrt{1 - \dfrac{4L}{R^2C}}\right] = -\dfrac{\omega_0}{2Q_s}\left[1 \pm \sqrt{1 - 4Q_s^2}\right]$

There are three cases. Consider the under damped case where Q>½.

(36) $\quad \mathscr{L}[i(t)] = \dfrac{V_m}{L}\dfrac{1}{(p+p_1)(p+p_2)} = \dfrac{V_m}{L}\dfrac{1}{p_1 - p_2}\left(\dfrac{1}{p+p_2} - \dfrac{1}{p+p_1}\right)$

$\quad = \dfrac{V_m}{L}\dfrac{1}{2j\gamma}\left(\dfrac{1}{p+\alpha - j\gamma} - \dfrac{1}{p+\alpha + j\gamma}\right) \quad \text{where } \beta = j\gamma$

(37) $\quad i(t) = \dfrac{V_m}{L}\dfrac{1}{2j\gamma}\left(e^{-(\alpha - j\gamma)t} - e^{-(\alpha + j\gamma)t}\right) = \dfrac{V_m}{L}\dfrac{1}{2j\gamma}e^{-\alpha t}\left(e^{j\gamma t} - e^{-j\gamma t}\right)$

$\quad = \dfrac{V_m}{\gamma L}e^{-\alpha t}\sin \gamma t$

---

[3] N. L. Pappas "Electric Circuits Analysis and Design" ISBN 978 1494 273 385

4.3 General Transforms

Transform of multiplication by a constant K
(3a) if $f(t) \Leftrightarrow F(p)$, then

$$\mathcal{L}[k \times f(t)] = \int_0^\infty k \times f(t)e^{-pt}\,dt = k \times \int_0^\infty f(t)e^{-pt}\,dt = k \times F(p)$$

(3b) $kf(t) \Leftrightarrow kF(p)$ (the $\Leftrightarrow$ symbol means transform from t to p or p to t)

Transform of a linear sum with constants c_1 and c_2
(4a) if $f_i(t) \Leftrightarrow F_i(p)$ $i = 1, 2$ then

$$\mathcal{L}[c_1 f_1(t) + c_2 f_2(t)] = \int_0^\infty (c_1 f_1(t) + c_2 f_2(t))e^{-pt}\,dt$$

$$= \int_0^\infty c_1 f_1(t)e^{-pt}\,dt + \int_0^\infty c_2 f_2(t)e^{-pt}\,dt = c_1 F_1(p) + c_2 F_2(p)$$

(4b) $c_1 f_1(t) + c_2 f_2(t) \Leftrightarrow c_1 F_1(p) + c_2 F_2(p)$

Transform of a first derivative with respect to variable t
(5a) if $f(t) \Leftrightarrow F(p)$ and $\dfrac{df(t)}{dt} \Leftrightarrow F_1(p)$, then

$$F_1(p) = \mathcal{L}\left[\frac{df(t)}{dt}\right] = \int_0^\infty \frac{df(t)}{dt} e^{-pt}\,dt$$

if $u = e^{-pt}$, $dv = \dfrac{df(t)}{dt}\,dt$ then $du = -pe^{-pt}\,dt$ $v = f(t)$

integrating by parts: $F_1(p) = \int_0^\infty u\,dv = uv\Big|_0^\infty - \int_0^\infty v\,du$

$$F_1(p) = e^{-pt} f(t)\Big|_0^\infty - \int_0^\infty f(t)(-p)e^{-pt}\,dt$$

$$= [e^{-p \times \infty} f(\infty) - e^{-p \times 0} f(0)] + p \int_0^\infty f(t)e^{-pt}\,dt$$

$$= [0 - f(0)] + pF(p) = pF(p) - f(0)$$

(5b) $\dfrac{df(t)}{dt} \Leftrightarrow pF(p) - f(0)$

Mathematics beyond the Calculus

Transform of higher derivatives with respect to variable t
(6) $\mathcal{L}[f''(t)] = p\{\mathcal{L}[f'(t)]\} - f'(0)$
$= p\{p\mathcal{L}[f(t)] - f(0)\} - f'(0)$
$= p^2 \mathcal{L}[f(t)] - pf(0) - f'(0)$
(7) $\mathcal{L}[f'''(t)] = p\{\mathcal{L}[f''(t)]\} - f''(0)$
$= p^3 \mathcal{L}[f(t)] - p^2 f(0) - pf'(0) - f''(0)$

Transform of an integral with respect to variable t

(8a) if $f(t) \Leftrightarrow F(p)$ and $\int_0^t f(x)dx \Leftrightarrow F_1(p)$, then

$$F_1(p) = \mathcal{L}[\int_0^t f(x)dx] = \int_0^\infty \int_0^t f(x)dx \, e^{-pt} dt$$

if $u = \int_0^t f(x)dx$ and $dv = e^{-pt}dt$ then $du = f(t)dt$ and $v = -\frac{1}{p}e^{-pt}$

$$F_1(p) = \int_0^\infty u \, dv = vu \Big|_0^\infty - \int_0^\infty v \, du$$

$$F_1(p) = \left[-\frac{1}{p}e^{-pt}\int_0^t f(x)dx\right]_0^\infty - \int_0^\infty \left(-\frac{1}{p}\right)f(t)e^{-pt}dt$$

$$= \left(-\frac{1}{p}e^{-p\infty}\int_0^\infty f(x)dx\right) - \left(-\frac{1}{p}e^{-p\cdot 0}\int_0^0 f(x)dx\right) + \frac{1}{p}\int_0^\infty f(t)e^{-pt}dt$$

$$= \left(-\frac{1}{p} \times 0 \int_0^\infty f(x)dx\right) - \left(-\frac{1}{p} \times 1 \times 0\right) + \frac{1}{p}F(p) = 0 - 0 + \frac{F(p)}{p}$$

(8b) $\int_0^t f(x)dx \Leftrightarrow \dfrac{F(p)}{p}$

Problem 401

$F(p) = \mathcal{L}[f(t)]$, ref 10c, show that $f(t) = \mathcal{L}^{-1}\left[\ln\dfrac{p+a}{p+b}\right] = \dfrac{1}{t}\left(e^{-bt} - e^{-at}\right)$

Problem 402 $F(p) = \mathcal{L}[f(t)]$, ref 11, show that $\mathcal{L}\left[\dfrac{1-e^{-t}}{t}\right] = \ln\left(1 + \dfrac{1}{p}\right)$

4 The Laplace Transform

Transform of a change in variable t to $t - c$

(9a) If $F(p) = \int_0^\infty f(t)e^{-pt}dt$, then let $F_c(p) = \int_0^\infty f(t-c)e^{-pt}dt$

let $\quad \tau = t - c$

$$F_c(p) = \int_0^\infty f(\tau)e^{-p(\tau+c)}d\tau = e^{-pc}\int_0^\infty f(\tau)e^{-p\tau}d\tau = e^{-pc}F(p)$$

If $f(t)u(t) \Leftrightarrow F(p)$ then

(9b) $f(t-c) \Leftrightarrow e^{-pc}F(p)$ \quad (9c) by a similar process $f(t+c) \Leftrightarrow e^{+pc}F(p)$

Derivatives of a transform $F(p)$

(10a) $\dfrac{dF(p)}{dp} = \dfrac{d}{dp}\int_0^\infty f(t)e^{-pt}dt = \int_0^\infty \dfrac{\partial}{\partial p}f(t)e^{-pt}dt = \int_0^\infty -t\,f(t)e^{-pt}dt$

$\qquad\qquad = \mathcal{L}[-t\,f(t)]$

(10b) $\dfrac{d^n F(p)}{dp^n} = \mathcal{L}[(-1)^n t^n f(t)] \qquad n = 1, 2, 3, \ldots$

(10c) When $n=1$ can rewrite 10a as follows $\quad f(t) = -\dfrac{1}{t}\mathcal{L}^{-1}\left[\dfrac{dF(p)}{dp}\right]$

Integral of a transform $F(p)$

(11a) $\quad F(x) = \int_0^\infty f(t)e^{-xt}dt \;\rightarrow\; \int_p^\infty F(x)dx = \lim_{w\to\infty}\int_p^w\left[\int_0^\infty f(t)e^{-xt}dt\right]dx \;(x\text{ real})$

$\int_p^\infty F(x)dx = \lim_{w\to\infty}\int_0^\infty\left[\int_p^w f(t)e^{-xt}dt\right]dx = \lim_{w\to\infty}\int_0^\infty\left[\dfrac{1}{-t}f(t)e^{-xt}\right]_p^w dx$

$\qquad = \lim_{w\to\infty}\int_0^\infty\left[\dfrac{1}{t}f(t)e^{-pt} - \dfrac{1}{t}f(t)e^{-wt}\right]dx = \int_0^\infty \dfrac{1}{t}f(t)e^{-pt}dt - 0$

(11b) $\int_p^\infty F(x)dx = \mathcal{L}\left[\dfrac{1}{t}f(t)\right]$

25

Mathematics beyond the Calculus

4.4 Specific Transforms

Many complicated functions of real variable t directly transform into elementary functions of a complex variable p.

Figure 401 Step V_m u(t)

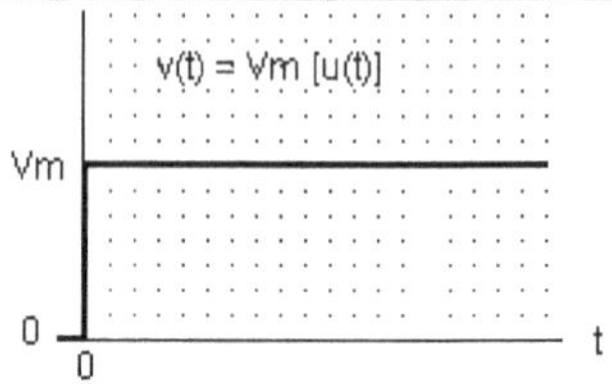

$v(t) = V_m [u(t)]$

Transform of the step function u(t) The unit step function (Figure 401,) transforms to the algebraic expression $1/p$.

(12a) $u(t) = 1 \quad for\ all\ t \geq 0$

$$F(p) = \mathcal{L}[u(t)] = \int_0^\infty u(t) e^{-pt} dt = \int_0^\infty 1 \times e^{-pt} dt = \left. \frac{e^{-pt}}{-p} \right|_0^\infty = \frac{0}{-p} - \frac{e^{-0}}{-p} = \frac{1}{p}$$

where $e^{-\infty} = 0$ means $\lim_{t \to \infty} e^{-pt} = 0$

(12b) $u(t) \Leftrightarrow \dfrac{1}{p}$

Transform of exp(–at) Most transcendental functions transform into algebraic expressions. Consider the exponential function.

(13a) $F(p) = \mathcal{L}[e^{-at}] = \int_0^\infty e^{-at} e^{-pt} dt = \int_0^\infty e^{-(p+a)t} dt = \left. \dfrac{e^{-(p+a)t}}{-(p+a)} \right|_0^\infty$

$= \dfrac{0}{-(p+a)} - \dfrac{e^{-0}}{-(p+a)} = \dfrac{1}{p+a}$

(13b) $e^{-at} \Leftrightarrow \dfrac{1}{p+a}$

The transform of u(t) follows from this result by letting a = 0. This is a useful method for generating other transform pairs.

Calculate the transforms. Hint – convert f(t) to exponential form.

Problem 403 (a) $\mathcal{L}[\sin at] = \dfrac{a}{p^2 + a^2}$ (b) $\mathcal{L}[\cos at] = \dfrac{p}{p^2 + a^2}$

Problem 404

(a) $\mathcal{L}[\sinh at] = \dfrac{a}{p^2 - a^2}$ (b) $\mathcal{L}[\cosh at] = \dfrac{p}{p^2 - a^2}$

4 The Laplace Transform

Transforms of damped sin ωt and cos ωt The sin and cos functions in exponential format reveal that the transform of an exponential can be used as a shortcut.

Damped sine function

(14a) $F(p) = \mathscr{L}[e^{-\sigma t}\sin\omega t]$

$$F(p) = \int_0^\infty \frac{e^{-(\sigma-i\omega)t} - e^{-(\sigma+i\omega)t}}{2i} e^{-pt}\,dt = \int_0^\infty \frac{e^{-(p+\sigma-i\omega)t} - e^{-(p+\sigma+i\omega)t}}{2i}\,dt$$

$$F(p) = \frac{1}{2i}\left(\frac{1}{p+\sigma-i\omega} - \frac{1}{p+\sigma+i\omega}\right) = \frac{\omega}{(p+\sigma)^2 + \omega^2}$$

(14b) $e^{-\sigma t}\sin\omega t \Leftrightarrow \dfrac{\omega}{(p+\sigma)^2 + \omega^2}$

Damped cosine function

(15a) $F(p) = \mathscr{L}[e^{-\sigma t}\cos\omega t]$

$$F(p) = \int_0^\infty \frac{e^{-(\sigma-i\omega)t} + e^{-(\sigma+i\omega)t}}{2} e^{-pt}\,dt = \int_0^\infty \frac{e^{-(p+\sigma-i\omega)t} + e^{-(p+\sigma+i\omega)t}}{2}\,dt$$

$$F(p) = \frac{1}{2}\left(\frac{1}{p+\sigma-i\omega} + \frac{1}{p+\sigma+i\omega}\right) = \frac{p+\sigma}{(p+\sigma)^2 + \omega^2}$$

(15b) $e^{-\sigma t}\cos\omega t \Leftrightarrow \dfrac{p+\sigma}{(p+\sigma)^2 + \omega^2}$

Transform of the ramp function We have no shortcuts for the ramp function integration. Here we integrate by parts.

(16a) ramp: $F(p) = \mathscr{L}[\dfrac{t}{T}] = \int_0^\infty \dfrac{t}{T} e^{-pt}\,dt = \dfrac{1}{T}\int_0^\infty t \times e^{-pt}\,dt = \dfrac{1}{T}\int_0^\infty u\,dv$

let $u = t,\ dv = e^{-pt}\,dt.\quad \Rightarrow \quad du = dt,\ v = -\dfrac{1}{p}e^{-pt}$

$$F(p) = \frac{1}{T}\int_0^\infty u\,dv = \frac{1}{T}uv\Big|_0^\infty - \frac{1}{T}\int_0^\infty v\,du$$

$$F(p) = -\frac{1}{T}t\frac{1}{p}e^{-pt}\Big|_0^\infty + \frac{1}{pT}\int_0^\infty e^{-pt}\,dt = 0 + \frac{1}{pT}\frac{-1}{p}e^{-pt}\Big|_0^\infty = \frac{1}{T}\frac{1}{p^2}$$

(16b) $t\,u(t) \Leftrightarrow \dfrac{1}{p^2}$

Mathematics beyond the Calculus

Calculate the transforms when $\mathcal{L}[f(y)] = F(p)$

Problem 405 Hint - equation (10)
$$\mathcal{L}[t\cos at] = \frac{p^2 - a^2}{(p^2 + a^2)^2}$$

Problem 406
$$\mathcal{L}[e^{bt}\sin at] = \frac{a}{(p-b)^2 + a^2}$$

Problem 407
$$\mathcal{L}\left[f\left(\frac{t}{a}\right)\right] = aF(ap)$$

Problem 408
$$\mathcal{L}[-t\,f(t)] = \frac{d}{dp}F(p)$$

Problem 409 Hint Integrate by parts
$$\mathcal{L}[4t] = \frac{4}{p^2}$$

Problem 410
$$\mathcal{L}[te^{2t}] = -\frac{1}{(p-2)^2}$$

Problem 411
$$\mathcal{L}[e^{ay}\cos by] = \frac{p-a}{(p-a)^2 + b^2}$$

Problem 412
$$\mathcal{L}[x\sin bx] = \frac{2bp}{(p^2 + b^2)^2}$$

Problem 413
$$\mathcal{L}[\sin at - at\cos at] = \frac{2a^3}{(p^2 + a^2)^2}$$

4 The Laplace Transform

4.5 Partial Fractions

We use a well known trick that allows us to avoid calculating formidable integral (2) for almost all problems we might ever encounter.

The trick is straightforward. Expand the algebraic solution into a sum of partial fractions. Then use the inverse transform of the exponential or other known transforms.

(17) $\mathscr{L}[e^{-at}] = \dfrac{1}{p+a} \quad \Rightarrow \quad e^{-at} \Leftrightarrow \dfrac{1}{p+a}$

The terms of partial fraction expansions of F(p) are transforms of exp(−at), because each term of a partial fraction expansion has a pole of order one or higher.

pole of order one   *pole of order n+1*

(18a) $f(t) = e^{-at} \Leftrightarrow F(p) = \dfrac{1}{p+a}$ (18b) $f(t) = \dfrac{t^n}{n!} e^{-at} \Leftrightarrow F(p) = \dfrac{1}{(p+a)^{n+1}}$

The sum of the inverse Laplace transforms of each term of the partial fraction equals the inverse Laplace transform of F(p). For example,

if $F(p) = \dfrac{17}{2j}\left(\dfrac{1}{p - j\omega_0} - \dfrac{1}{p + j\omega_0}\right) + \dfrac{1}{p + \dfrac{R}{L}} + \dfrac{8}{p}$ then

(19) $f(t) = \left[\dfrac{17}{2j}\left(e^{j\omega_0 t} - e^{-j\omega_0 t}\right) + e^{-\frac{R}{L}t} + 8\right] u(t) = \left(17\sin\omega_0 t + e^{-\frac{R}{L}t} + 8\right) u(t)$

Any complex frequency domain function arising from the linear, lumped parameter electric circuits is the ratio of two polynomials in the complex frequency variable p. We can always divide the denominator into the numerator so that the remainder F(p) is a proper rational fraction that means the degree of N(p) is less than the degree of D(p). For example:

if $G(p) = \dfrac{p^4 + 5p^3 + 8p^2 + 3p + 6}{p^3 + 4p^2 + 3p} = p + 1 + \dfrac{p^2 + 6}{p^3 + 4p^2 + 3p}$

(20) $\text{let } F(p) = \dfrac{p^2 + 6}{p^3 + 4p^2 + 3p} = \dfrac{N(p)}{D(p)}$ (a proper fraction)

> Note: $f(t) = \dfrac{d\delta(t)}{dt} \Leftrightarrow p$ and $f(t) = \delta(t) \Leftrightarrow 1$
> where $\delta(t)$ is the impulse function

Mathematics beyond the Calculus

There are two problems to solve:
1. Find the roots of D(p) (we assume you know how to do this).
2. Convert F(p) into a sum of terms with known inverse Laplace transforms.

1. Method for any roots - equate coefficients This method is based on the fact coefficients of similar terms of two polynomials are equal.

(21) $\quad F(p) = \dfrac{p^2 + 6}{p^3 + 4p^2 + 3p}$

(22) $\quad \dfrac{p^2 + 6}{p(p+1)(p+3)} = \dfrac{k_1}{p} + \dfrac{k_2}{p+1} + \dfrac{k_3}{p+3}$

cross multiply :

(23) $\quad p^2 + 6 = k_1(p+1)(p+3) + k_2 p(p+3) + k_3 p(p+1)$

Do not simplify by multiplying out. It is easier to substitute selected values of p.

(24a) *if* $p = 0$ *then*
$$0 + 6 = k_1(0+1)(0+3) + k_2 0(0+3) + k_3 0(0+1)$$
$$6 = 3k_1 \quad \Rightarrow \quad (8) \quad k_1 = 2$$

(24b) *if* $p = -1$ *then*
$$1 + 6 = k_1(-1+1)(-1+3) + k_2(-1)(-1+3) + k_3(-1)(-1+1)$$
$$7 = -2k_2 \quad \Rightarrow \quad (9) \quad k_2 = -\dfrac{7}{2}$$

(24c) *if* $p = -3$ *then*
$$9 + 6 = k_1(-3+1)(-3+3) + k_2(-3)(-3+3) + k_3(-3)(-3+1)$$
$$15 = 6k_3 \quad \Rightarrow \quad (10) \quad k_3 = \dfrac{5}{2}$$

Higher order roots Each higher order root of order n requires a sum of terms. One term for each power of p from 1 to n.

(25) $\quad F(p) = \dfrac{5p^3 - 6p - 3}{p^3(p+1)^2}$

(26) $\quad \dfrac{5p^3 - 6p - 3}{p^3(p+1)^2} = \dfrac{k_1}{p} + \dfrac{k_2}{p^2} + \dfrac{k_3}{p^3} + \dfrac{k_4}{p+1} + \dfrac{k_5}{(p+1)^2}$ *now cross multiply*

(27) $\quad 5p^3 - 6p - 3 = k_1 p^2(p+1)^2 + k_2 p(p+1)^2 + k_3(p+1)^2 + k_4 p^3(p+1) + k_5 p^3$

if $p = 0$, *then* $-3 = k_3$, *and if* $p = -1$, *then* $-5 + 6 - 3 = -k_5$

4 The Laplace Transform

If we equate coefficients of terms, then for

$p^4: 0 = k_1 + k_4 \qquad p^3: 5 = 2k_1 + k_2 + k_4 + k_5 \qquad p: -6 = k_2 + 2k_3$

(28) $\quad k_3 = -3, \ k_5 = 2, \ k_2 = 0, \ k_1 = 3, \ k_4 = -3$

Quadratic factors Quadratic factors require numerators to be linear in p.

(29) $\qquad F(p) = \dfrac{16}{p(p^2 + 4)^2}$

(30) $\qquad \dfrac{16}{p(p^2+4)^2} = \dfrac{k_1}{p} + \dfrac{k_2 p + k_3}{p^2 + 4} + \dfrac{k_4 p + k_5}{(p^2+4)^2}$

Now cross multiply

$16 = k_1(p^2 + 4)^2 + (k_2 p + k_3)p(p^2 + 4) + (k_4 p + k_5)p$

(31) $\quad 16 = k_1(p^4 + 8p^2 + 16) + k_2(p^4 + 4p^2) + k_3(p^3 + 4p) + k_4 p^2 + k_5 p$

if $p = 0$, then $16 = k_1 16 \quad k_1 = 1$

If we equate coefficients of terms, then for

$p^4) \ 0 = k_1 + k_2 \qquad p^3) \ 0 = k_3 \qquad p^2) \ 0 = 8k_1 + 4k_2 + k_4 \qquad p) \ 0 = 4k_3 + k_5$

(32) $\quad k_1 = 1, \ k_2 = -1, \ k_3 = 0, \ k_4 = -4, \ k_5 = 0$

2. Method for real roots of order 1 This is a formal equivalent to cross-multiplying. The process is straightforward when roots are real and distinct.

(33) $\qquad F(p) = \dfrac{14p + 11}{(p+1)(p+3)}$

(34) $\qquad \dfrac{14p + 11}{(p+1)(p+3)} = \dfrac{k_1}{p+1} + \dfrac{k_2}{p+3}$

(35) $\qquad k_1 = (p+1)F(p)\big|_{p=-1} = \dfrac{14(-1)+11}{(-1+3)} = -\dfrac{3}{2}$

(36) $\qquad k_2 = (p+3)F(p)\big|_{p=-3} = \dfrac{14(-3)+11}{(-3+1)} = \dfrac{31}{2}$

(37) $\qquad F(p) = -\dfrac{3}{2}\dfrac{1}{p+1} + \dfrac{31}{2}\dfrac{1}{p+3} \Rightarrow f(t) = -\dfrac{3}{2}e^{-t} + \dfrac{31}{2}e^{-3t}$

Mathematics beyond the Calculus

Use the trick (page 29) to compute the inverse transform.

Problem 414

show that $f(t) = \mathscr{L}^{-1}\left[\dfrac{p+1}{p^2(p-1)}\right] = -2 - t + 2e^t$

Problem 415

show that $f(t) = \mathscr{L}^{-1}\left[\dfrac{2p^2}{(p^2+1)(p-1)^2}\right] = -\cos t + e^t + te^t$

Problem 416

show that $f(t) = \mathscr{L}^{-1}\left[\dfrac{p}{(p+2)(p-1)(p-3)}\right] = -\dfrac{1}{6}e^t - \dfrac{2}{15}e^{-2t} + \dfrac{3}{10}e^{3t}$

Problem 417

show that $f(t) = \mathscr{L}^{-1}\left[\dfrac{p^2+1}{p(p-1)^2}\right] = -1 + e^t + t^2 e^t$

Problem 418

show that $f(t) = \mathscr{L}^{-1}\left[\dfrac{2p^2+3}{(p+1)^2(p^2+1)}\right] = \dfrac{3}{2}e^{-t} + \dfrac{5}{4}te^{-t} - \dfrac{3}{2}\cos t + \dfrac{1}{4}\sin t - \dfrac{1}{4}t\sin t$

Problem 419

show that $f(t) = \mathscr{L}^{-1}\left[\dfrac{p+2}{p^5 - 3p^4 + 2p^3}\right] = \dfrac{5}{2} + 2t + \dfrac{t^2}{2} - 3e^t + \dfrac{1}{2}e^{2t}$

Problem 420

show that $f(t) = \mathscr{L}^{-1}\left[\dfrac{p^3}{p^4 + 4\omega^4}\right] = \cos \omega t \cosh \omega t$

Problem 421

show that $f(t) = \mathscr{L}^{-1}\left[\dfrac{\Gamma(n+1)}{p^{n+1}}\right] = x^n \quad n > -1$

4 The Laplace Transform

4.6 Periodic Functions

If a function f(t) is periodic with period T, then f(t) = f(t+T) for all t so that the transform of a periodic function is as follows.

$$(38a) \quad \mathcal{L}[f(t)] = \int_0^\infty f(t)e^{-pt}\,dt = \int_0^T f(t)e^{-pt}\,dt + \int_T^{2T} f(t)e^{-pt}\,dt + \int_{2T}^{3T} f(t)e^{-pt}\,dt + \ldots$$

$$(38b) \quad \mathcal{L}[f(t)] = \int_0^T f(t)e^{-pt}\,dt + \int_0^T f(t)e^{-p(t+T)}\,dt + \int_0^T f(t)e^{-p(t+2T)}\,dt + \ldots$$

$$(38c) \quad \mathcal{L}[f(t)] = \int_0^T f(t)e^{-pt}\{1 + e^{-pT} + e^{-2pT} + \ldots\}\,dt$$

$$(38d) \quad \mathcal{L}[f(t)] = \frac{1}{1-e^{-pT}} \int_0^T f(t)e^{-pt}\,dt$$

For example:
A squarewave = 0 for t from 0 to T/2 and it = 1 from T/2 to T.

$$(39) \quad \int_0^T f(t)e^{-pt}\,dt = \int_{T/2}^T 1 \cdot e^{-pt}\,dt = \left.\frac{e^{-pt}}{-p}\right|_{T/2}^T = \frac{e^{-pT/2} - e^{-pT}}{p}$$

$$\mathcal{L}[f(t)] = \frac{1}{1-e^{-pT}} \int_0^T f(t)e^{-pt}\,dt = \frac{1}{1-e^{-pT}} \cdot \frac{e^{-pT/2} - e^{-pT}}{p} = \frac{e^{-pT/2}(1-e^{-pT/2})}{p(1-e^{-pT})}$$

note that $(1-e^{-pT}) = (1-e^{-pT/2})(1+e^{-pT/2})$ so that

$$\mathcal{L}[f(t)] = \frac{e^{-pT/2}}{p(1+e^{-pT/2})} = \frac{1}{p(1+e^{pT/2})}$$

Problem 422
A squarewave = 1 for t from 0 to T/2 and it = −1 from T/2 to T.
show that the periodic transform of the ± 1 squarewave is

$$\mathcal{L}[f(t)] = \frac{1}{p} \times \frac{(1-e^{-pT/2})}{(1+e^{-pT/2})}$$

4.7 The Gamma Function

Euler's Gamma Function is useful in evaluating certain definite integrals, which arise in physical problems. It is a generalization of the factorial function, which is written as n! where
(40a) $n! = n(n-1)(n-2)(n-3)......$ → for example $5! = 5 \times 4 \times 3 \times 2 \times 1$
(40b) clearly $(n)! = n \cdot (n-1)!$

Factorial n only has integer values 1, 2, 6, 24, etc, which produce a discontinuous plot. Euler's Gamma Function generalization produces a continuous plot that fills in the gaps between the integer values of n!

Our guess is that Euler deduced he needed an integral with a property analogous to equation 41.

(41) $$\int_0^\infty f_n(x)dx = n \int_0^\infty f_{n-1}(x)dx$$

Probably sooner than later, Euler found the solution.

(42a) integrate by parts $\int_0^\infty x^n e^{-x} dx$

where $u = x^n \quad du = nx^{n-1} \quad dv = e^{-x}dx \quad v = -e^{-x}$

(42b) $$\int_0^\infty x^n e^{-x} dx = -x^n e^{-x} \Big|_0^\infty + n \int_0^\infty x^{n-1} e^{-x} dx$$

(42c) $$\int_0^\infty x^n e^{-x} dx = n \int_0^\infty x^{n-1} e^{-x} dx \quad \rightarrow \quad f_n = n \cdot f_{n-1}$$

The Gamma function is defined for all real values of n *except 0 and negative integers*.

(43) $\Gamma(n) = \int_0^\infty x^{n-1} e^{-x} dx \quad (n > 0) \quad$ and $\quad \Gamma(n+1) = n\Gamma(n)$

Emphasis - Observe that $\Gamma(n)$ is analogous to $(n-1)!$ not n!

FYI – Pierce's Tables include a table of numerical values of $\Gamma(n)$. Also see Abramowitz, M. et al "Handbook of Mathematical Functions"

Laplace Transforms involving Γ and γ

(44a) $\quad \mathcal{L}[t^v] = \int_0^\infty t^v e^{-pt} dt \quad (v \geq 0)$

(44b) $\quad$ let $x = pt \quad \rightarrow \quad \mathcal{L}[t^v] = \int_0^\infty \left(\frac{x}{p}\right)^v e^{-x} \frac{1}{p} dx = \frac{1}{p^{v+1}} \int_0^\infty x^v e^{-x} dx$

(44c) $\quad \mathcal{L}[t^v] = \frac{1}{p^{v+1}} \Gamma(v+1)$

(45a) $\quad \mathcal{L}[\ln t] = \int_0^\infty \ln t \cdot e^{-pt} dt$

(45b) $\quad$ let $x = pt \quad \rightarrow \quad \mathcal{L}[\ln t] = \int_0^\infty \ln \frac{x}{p} e^{-x} \frac{1}{p} dx$

$\mathcal{L}[\ln t] = \frac{1}{p}\left(\int_0^\infty \ln x \cdot e^{-x} dx - \ln p \int_0^\infty e^{-x} dx\right) = -\frac{1}{p}(-\gamma - \ln p \cdot 1)$

(45c) $\quad \mathcal{L}[\ln t] = \frac{1}{p}(\ln p + \gamma) \quad (\gamma = 0.577215...)$

4.8 More Properties of Transforms

Change of Scale

(46a) $\quad$ if $F(p) = \mathcal{L}[f(\tau)] = \int_0^\infty f(\tau) e^{-\tau p} d\tau \quad$ then

(46b) $\quad$ the exponent $-\tau p$ of $e^{-\tau p}$ is changed to $-st$ as follows

$\quad$ let $\tau = \frac{t}{a}$ and s be a complex variable analogous to p so that

(46c) $\quad \tau p = \frac{t}{a} p = ts \quad \rightarrow \quad p = sa \quad \rightarrow \quad F(p) = \mathcal{L}[f(\tau)] \quad \rightarrow \quad F(as) = \mathcal{L}[f(\frac{t}{a})]$

(46d) $\quad F(as) = \int_0^\infty f\left(\frac{t}{a}\right) e^{-\left(\frac{t}{a}\right)p}\left(\frac{1}{a}\right) dt = \left(\frac{1}{a}\right) \int_0^\infty f\left(\frac{t}{a}\right) e^{-st} dt = \left(\frac{1}{a}\right) \mathcal{L}[f(\frac{t}{a})]$

(46e) $\quad F(as) = \left(\frac{1}{a}\right) \mathcal{L}[f(\frac{t}{a})] \quad$ or $\quad aF(ap) = \mathcal{L}[f(\frac{t}{a})]$

Example 1

(47a) given this transform pair where t is in seconds

$$\frac{p+0.5}{(p+0.5)^2 + \pi^2} \Leftrightarrow e^{-0.5t} \cos \pi t \quad t \geq 0$$

(47b) scale the pair so that t is in 0.5 seconds i.e. $a = 2$

$$\frac{p+0.25}{(p+0.25)^2 + (\pi/2)^2} \Leftrightarrow e^{-0.25t} \cos \pi \frac{t}{2} \quad t \geq 0$$

35

Mathematics beyond the Calculus

Example 2

(48a) if $f(t) = \mathcal{L}^{-1}[F(p)] = \mathcal{L}^{-1}\left[\dfrac{10^6}{(p+0.02\times 10^6)(p+0.8\times 10^6)}\right]$

(48b) then $f\left(\dfrac{t}{a}\right) = \mathcal{L}^{-1}[aF(ap)]$ so that $f\left(\dfrac{t}{10^6}\right) = \mathcal{L}^{-1}[10^6 F(10^6 p)]$

$= \mathcal{L}^{-1}\left[10^6 \dfrac{10^6}{(10^6 p + 0.02\times 10^6)(10^6 p + 0.8\times 10^6)}\right] = \mathcal{L}^{-1}\left[\dfrac{1}{(p+0.02)(p+0.8)}\right]$

(48c) $f\left(\dfrac{t}{10^6}\right) = \dfrac{e^{-0.02t} - e^{-0.8t}}{0.78} \rightarrow f(t) = f\left(10^6 \dfrac{t}{10^6}\right) = \dfrac{e^{-0.02\times 10^6 t} - e^{-0.8\times 10^6 t}}{0.78}$

Complex Integration

(66) if $\mathcal{L}[f(t)] = \displaystyle\int_0^\infty f(t)e^{-pt}\,dt = F(p)$ then

$$\int_p^\infty F(p)\,dp = \int_p^\infty \int_0^\infty f(t)e^{-pt}\,dt\,dp = \int_0^\infty f(t) \int_p^\infty e^{-pt}\,dp\,dt = \int_0^\infty f(t)\dfrac{1}{t}e^{-pt}\,dt$$

therefore $\mathcal{L}\left[\dfrac{f(t)}{t}\right] = \displaystyle\int_p^\infty F(p)\,dp$

Examples

(67) $\dfrac{2ap}{(p^2+a^2)^2} \Leftrightarrow t\sin at \rightarrow \dfrac{f(t)}{t} = \dfrac{t\sin at}{t} = \sin at$

$\displaystyle\int_p^\infty F(p)\,dp = \int_p^\infty \dfrac{a}{(p^2+a^2)^2}\,2p\,dp = \left.\dfrac{-a}{(p^2+a^2)}\right|_p^\infty = \dfrac{a}{(p^2+a^2)}$

therefore $\dfrac{a}{(p^2+a^2)} \Leftrightarrow \sin at$

(68) $\dfrac{1}{p+a} - \dfrac{1}{p+b} \Leftrightarrow e^{-at} - e^{-bt} \rightarrow \dfrac{f(t)}{t} = \dfrac{e^{-at} - e^{-bt}}{t}$

$\displaystyle\int_p^\infty F(p)\,dp = \int_p^\infty \left(\dfrac{1}{p+a} - \dfrac{1}{p+b}\right)dp = \left.\dfrac{\ln(p+a)}{\ln(p+b)}\right|_p^\infty = 0 - \dfrac{\ln(p+a)}{\ln(p+b)} = \dfrac{\ln(p+b)}{\ln(p+a)}$

therefore $\dfrac{\ln(p+b)}{\ln(p+a)} \Leftrightarrow \dfrac{e^{-at} - e^{-bt}}{t}$

4 The Laplace Transform

Complex Multiplication - Real Convolution

Sometimes the inverse transform of transform F(p) is easier to accomplish if F(p) is treated as the product of F(p)'s. I.e.

(49) $f(t) = \mathscr{L}^{-1}\left[\dfrac{2p^2}{(p^2+1)(p-1)^2}\right] = \mathscr{L}^{-1}\left[\dfrac{2p^2}{(p^2+1)} \times \dfrac{1}{(p-1)^2}\right]$
$= \mathscr{L}^{-1}[F_1(p) \times F_2(p)]$

The product is a complex multiplication whose inverse is expressed as a convolution integral.

(50) $\int_0^t f_1(t-\tau)f_2(\tau)d\tau = \int_0^t f_1(\tau)f_2(t-\tau)d\tau = \mathscr{L}^{-1}[F_1(p) \times F_2(p)]$

where $f_1(t) = \mathscr{L}^{-1}[F_1(p)]$ and $f_2(t) = \mathscr{L}^{-1}[F_2(p)]$

Examples
First just calculate the convolution integral.

(51) if $f_1(t) = e^t$ and $f_2(t) = t$

then $\int_0^t f_1(\tau)f_2(t-\tau)d\tau = \int_0^t e^\tau(t-\tau)d\tau = te^\tau\Big|_0^t - (\tau e^\tau - e^\tau)\Big|_0^t = e^t - t - 1$

Then start from transforms F(p)

(52) $\mathscr{L}^{-1}\left[\dfrac{1}{p^2} \times \dfrac{1}{(p-1)}\right] = \int_0^t e^\tau(t-\tau)d\tau = te^\tau\Big|_0^t - (\tau e^\tau - e^\tau)\Big|_0^t = e^t - t - 1$

Another example

(53a) find $\int_0^t f_1(t-\tau)f_2(\tau)d\tau = \mathscr{L}^{-1}\left[\dfrac{1}{p+a} \times \dfrac{1}{(p+b)^2}\right]$

where $e^{-at} = \mathscr{L}^{-1}\left[\dfrac{1}{p+a}\right]$ and $te^{-bt} = \mathscr{L}^{-1}\left[\dfrac{1}{(p+b)^2}\right]$

(53b) $\mathscr{L}^{-1}\left[\dfrac{1}{p+a} \times \dfrac{1}{(p+b)^2}\right] = \int_0^t e^{-a(t-\tau)}\tau e^{-b\tau}d\tau = e^{-at}\int_0^t e^{a\tau}\tau e^{-b\tau}d\tau$

$= e^{-at}\int_0^t \tau e^{(a-b)\tau}d\tau = \dfrac{e^{-at}}{(a-b)^2} + \dfrac{[(a-b)t-1]e^{-bt}}{(a-b)^2}$

Mathematics beyond the Calculus

Complex Translation

(54a) if $\mathscr{L}[f(t)] = F(p)$
then $\mathscr{L}[e^{-at} f(t)] = F(p+a)$ and $\mathscr{L}[e^{at} f(t)] = F(p-a)$

These results follow immediately from

(54b) $\int_0^\infty f(t)e^{-pt} dt = F(p)$

$\rightarrow \int_0^\infty [e^{-at} f(t)]e^{-pt} dt = \int_0^\infty [f(t)]e^{-(p+a)t} dt = F(p+a)$

and $\int_0^\infty [e^{at} f(t)]e^{-pt} dt = \int_0^\infty [f(t)]e^{-(p-a)t} dt = F(p-a)$

Examples

(55) $\mathscr{L}[\cos at] = \dfrac{p}{p^2 + a^2}$ then $\mathscr{L}[e^{-at} \cos at] = \dfrac{p+a}{(p+a)^2 + a^2}$

Differentiation with respect to a second variable a in f(a, t).

(56) if $\mathscr{L}[f(a,t)] = F(a,p)$ then $\mathscr{L}_t \left[\dfrac{\partial}{\partial a} f(a,t) \right] = \dfrac{\partial}{\partial a} F(a,p)$

Examples

(57) if $\mathscr{L}[\sin bt] = \dfrac{b}{p^2 + b^2}$ then $\dfrac{\partial}{\partial b} \sin bt = t \cos bt$ and

$\dfrac{\partial}{\partial b} \dfrac{b}{p^2 + b^2} = \dfrac{1}{(p^2 + b^2)} + \dfrac{b(-2b)}{(p^2 + b^2)^2} = \dfrac{p^2 - b^2}{(p^2 + b^2)^2}$

therefore $t \cos bt \Leftrightarrow \dfrac{p^2 - b^2}{(p^2 + b^2)^2}$

(58) if $\mathscr{L}[e^{-at} \cos bt] = \dfrac{p+a}{(p+a)^2 + b^2}$ then $\dfrac{\partial}{\partial a} e^{-at} \cos bt = -te^{-at} \cos bt$ and

$\dfrac{\partial}{\partial a} \dfrac{p+a}{(p+a)^2 + b^2} = (p+a) \dfrac{-1 \times 2(p+a)}{[(p+a)^2 + b^2]^2} + \dfrac{1}{(p+a)^2 + b^2} = -\dfrac{(p+a)^2 - b^2}{[(p+a)^2 + b^2]^2}$

therefore $e^{-at} \cos bt \Leftrightarrow \dfrac{(p+a)^2 - b^2}{[(p+a)^2 + b^2]^2}$

4 The Laplace Transform

Second Independent Variable

(59) if $\mathscr{L}[f(a,t)] = F(a,p)$ then $\mathscr{L}\left[\lim_{a \to a_0} f(a,t)\right] = \lim_{a \to a_0} F(a,p)$

The limit process is invariant under the transformation.

Example

(53b) $\dfrac{1}{p+a} \times \dfrac{1}{(p+b)^2} \Leftrightarrow \dfrac{e^{-at}}{(a-b)^2} + \dfrac{[(a-b)t-1]e^{-bt}}{(a-b)^2}$

The transform pair produces other pairs if $a \to 0$ and/or $b \to 0$

(60) $a \to 0$ then $\dfrac{1}{p} \times \dfrac{1}{(p+b)^2} \Leftrightarrow \dfrac{1}{b^2} + \dfrac{[(-b)t-1]e^{-bt}}{(-b)^2} = \dfrac{1-(bt+1)e^{-bt}}{b^2}$

(61) $b \to 0$ then $\dfrac{1}{p+a} \times \dfrac{1}{(p)^2} \Leftrightarrow \dfrac{e^{-at}}{(a)^2} + \dfrac{[(a)t-1]}{(a)^2} = \dfrac{e^{-at}+at-1}{a^2}$

(62) in (61) if $a \to 0$ then $\dfrac{e^{-at}+at-1}{a^2} = \dfrac{1+0-1}{0} = \dfrac{0}{0}$

this rquires use of l' Hospital's rule two times because

$\dfrac{d/da}{d/da} \dfrac{e^{-at}+at-1}{a^2} = \dfrac{-te^{-at}+t-0}{2a}$ and the limit is again $\dfrac{0}{0}$

$\dfrac{d/da}{d/da} \dfrac{-te^{-at}+t-0}{2a} = \dfrac{t^2 e^{-at}}{2}$ and the limit is $\dfrac{t^2}{2}$

therefore $\dfrac{1}{p^3} \Leftrightarrow \dfrac{t^2}{2}$

(63) in (56b) if $a \to b$ then $\dfrac{e^{-at}+[(a-b)t-1]e^{-bt}}{(a-b)^2} = \dfrac{0}{0}$

this rquires use of l' Hospital's rule two time because

$\dfrac{d/da}{d/da} \dfrac{e^{-at}+[(a-b)t-1]e^{-bt}}{(a-b)^2} = \dfrac{-te^{-at}+[(-b)t-0]e^{-bt}}{2(a-b)} = \dfrac{-te^{-at}-bte^{-bt}}{2(a-b)}$

$\dfrac{d/da}{d/da} \dfrac{-te^{-at}-bte^{-bt}}{2(a-b)} = \dfrac{t^2 e^{-at}-0}{2} = \dfrac{t^2 e^{-at}}{2}$ and the limit is $\dfrac{t^2 e^{-bt}}{2}$

therefore $\dfrac{1}{(p+b)^3} \Leftrightarrow \dfrac{t^2 e^{-bt}}{2}$

Mathematics beyond the Calculus

Integration with respect to a second variable a in f(a, t).

(69) if $\mathcal{L}[f(a,t)] = \int_0^\infty f(a,t)e^{-pt}\,dt = F(a,p)$ then

$$\int_{a_0}^{a} F(a,p)\,da = \int_{a_0}^{a}\int_0^\infty f(a,t)e^{-pt}\,dt\,da = \int_0^\infty \left(\int_{a_0}^{a} f(a,t)\,da\right) e^{-pt}\,dt$$

therefore $\mathcal{L}\left[\int_{a_0}^{a} f(a,t)\,da\right] = \int_{a_0}^{a} F(a,p)\,da$

Example

(70) $\dfrac{p}{p^2 + a^2} \Leftrightarrow \cos at \rightarrow \int_0^a \cos at\,da = \left[\dfrac{\sin at}{t}\right]_0^a = \dfrac{\sin at}{t}$

$\int_0^a \dfrac{p}{p^2 + a^2}\,da = \left[\arctan\dfrac{a}{p}\right]_0^a = \arctan\dfrac{a}{p}$

therefore $\arctan\dfrac{a}{p} \Leftrightarrow \dfrac{\sin at}{t}$

Real Multiplication - Complex Convolution

(2) $f(t) = \mathcal{L}^{-1}[F(p)] = \dfrac{1}{2\pi i}\int_{\sigma - j\infty}^{\sigma + j\infty} F(p)e^{tp}\,dp$

Applying (2) we get

(71) let $F(p) = \mathcal{L}[f_1(t) \times f_2(t)] = \int_0^\infty f_1(t) \times f_2(t) \times e^{-pt}\,dt$ then

$$F(p) = \int_0^\infty f_1(t) \times \dfrac{1}{2\pi i}\int_{\sigma - j\infty}^{\sigma + j\infty} F_2(w)e^{tw}\,dw \times e^{-pt}\,dt$$

$$= \dfrac{1}{2\pi i}\int_{\sigma - j\infty}^{\sigma + j\infty} F_2(w) \times \int_0^\infty f_1(t)e^{-pt}\,dt \times e^{tw}\,dw$$

$$= \dfrac{1}{2\pi i}\int_{\sigma - j\infty}^{\sigma + j\infty} F_2(w) \times \int_0^\infty f_1(t)e^{-(p-w)t}\,dt \times dw$$

$$= \dfrac{1}{2\pi i}\int_{\sigma - j\infty}^{\sigma + j\infty} F_2(w) \times F_1(p - w)\,dw$$

therefore $\mathcal{L}[f_1(t) \times f_2(t)] = \dfrac{1}{2\pi i}\int_{\sigma - j\infty}^{\sigma + j\infty} F_1(p - w) \times F_2(w)\,dw$

General Transforms

(3) $\mathscr{L}[kf(t)] = kF(p)$

(4) $\mathscr{L}[c_1 f_1(t) + c_2 f_2(t)] = c_1 F_1(p) + c_2 F_2(p)$

(5) $\mathscr{L}\left[\dfrac{df(t)}{dt}\right] = \mathscr{L}[f'(t)] = pF(p) - f(0)$

(6) $\mathscr{L}[f''(t)] = p^2 \mathscr{L}[f(t)] - pf(0) - f'(0)$

(7) $\mathscr{L}[f'''(t)] = p^3 \mathscr{L}[f(t)] - p^2 f(0) - pf'(0) - f''(0)$

(8) $\mathscr{L}\left[\displaystyle\int_0^t f(x)dx\right] = \dfrac{F(p)}{p}$

(9) $\mathscr{L}[f(t-c)] = e^{-pc} F(p)$ and $\mathscr{L}[f(t+c)] = e^{+pc} F(p)$

(10) $\dfrac{d^n F(p)}{dp^n} = \mathscr{L}[(-1)^n t^n f(t)] \qquad n = 1, 2, 3, \ldots$

(11) $\displaystyle\int_p^\infty F(x)dx = \mathscr{L}\left[\dfrac{1}{t} f(t)\right]$

(46e) $F(as) = \left(\dfrac{1}{a}\right)\mathscr{L}[f(\tfrac{t}{a})]$ or $aF(ap) = \mathscr{L}[f(\tfrac{t}{a})]$

(50) $\displaystyle\int_0^t f_1(t-\tau) f_2(\tau) d\tau = \int_0^t f_1(\tau) f_2(t-\tau) d\tau = \mathscr{L}^{-1}[F_1(p) \times F_2(p)]$
where $f_1(t) = \mathscr{L}^{-1}[F_1(p)]$ and $f_2(t) = \mathscr{L}^{-1}[F_2(p)]$

(54a) $\mathscr{L}[e^{-at} f(t)] = F(p+a)$ and $\mathscr{L}[e^{at} f(t)] = F(p-a)$

(56) if $\mathscr{L}[f(a,t)] = F(a,p)$ then $\mathscr{L}_t\left[\dfrac{\partial}{\partial a} f(a,t)\right] = \dfrac{\partial}{\partial a} F(a,p)$

Mathematics beyond the Calculus

(59) if $\mathscr{L}[f(a,t)] = F(a,p)$ then $\mathscr{L}\left[\lim_{a \to a_0} f(a,t)\right] = \lim_{a \to a_0} F(a,p)$

(64) $\mathscr{L}[t f(t)] = -\dfrac{d}{dp} F(p)$

(66) $\mathscr{L}\left[\dfrac{f(t)}{t}\right] = \int_p^\infty F(p) dp$

(69) $\mathscr{L}\left[\int_{a_0}^{a} f(a,t) da\right] = \int_{a_0}^{a} F(a,p) da$

(71) $\mathscr{L}[f_1(t) \times f_2(t)] = \dfrac{1}{2\pi i} \int_{\sigma - j\infty}^{\sigma + j\infty} F_1(p-w) \times F_2(w) \times dw$

Specific Transforms

$u(t) \Leftrightarrow \dfrac{1}{p}$ or $1 \Leftrightarrow \dfrac{1}{p}$

$\delta(t) \Leftrightarrow 1 \qquad \dfrac{d\delta(t)}{dt} \Leftrightarrow p \qquad \delta(t-a) \Leftrightarrow e^{-ap} \qquad \dfrac{d\delta(t-a)}{dt} \Leftrightarrow p e^{-ap}$

$\dfrac{t^n}{n!} \Leftrightarrow \dfrac{1}{p^{n+1}} \quad \to \quad t \Leftrightarrow \dfrac{1}{p^2}$

$e^{-at} \Leftrightarrow \dfrac{1}{p+a}$

$\dfrac{1}{a}(e^{at} - 1) \Leftrightarrow \dfrac{1}{p(p-a)}$

$\dfrac{e^{at} - e^{bt}}{a - b} \Leftrightarrow \dfrac{1}{(p-a)(p-b)} \qquad a \ne b$

$\dfrac{a e^{at} - b e^{bt}}{a - b} \Leftrightarrow \dfrac{p}{(p-a)(p-b)} \qquad a \ne b$

4 The Laplace Transform

$$(1+at)e^{at} \Leftrightarrow \frac{p}{(p-a)^2}$$

$$e^{-\sigma t}\sin\omega t \Leftrightarrow \frac{\omega}{(p+\sigma)^2+\omega^2} \quad \rightarrow \quad \sin\omega t \Leftrightarrow \frac{\omega}{p^2+\omega^2}$$

$$e^{-\sigma t}\cos\omega t \Leftrightarrow \frac{p+\sigma}{(p+\sigma)^2+\omega^2} \quad \rightarrow \quad \cos\omega t \Leftrightarrow \frac{p}{p^2+\omega^2}$$

$$t\cos at \Leftrightarrow \frac{p^2-a^2}{(p^2+a^2)^2} \qquad t\sin at \Leftrightarrow \frac{2ap}{(p^2+a^2)^2}$$

$$e^{-\sigma t}\sinh\omega t \Leftrightarrow \frac{\omega}{(p-\sigma)^2+\omega^2} \quad \rightarrow \quad \sinh\omega t \Leftrightarrow \frac{\omega}{p^2-\omega^2}$$

$$e^{-\sigma t}\cosh\omega t \Leftrightarrow \frac{p+\sigma}{(p-\sigma)^2+\omega^2} \quad \rightarrow \quad \cosh\omega t \Leftrightarrow \frac{p}{p^2-\omega^2}$$

$$t\cosh\omega t \Leftrightarrow \frac{p^2+\omega^2}{(p^2-\omega^2)^2} \quad \text{and} \quad t\sinh\omega t \Leftrightarrow \frac{2\omega p}{(p^2-\omega^2)^2}$$

$$\frac{1}{\sqrt{\pi t}} \Leftrightarrow \frac{1}{\sqrt{p}}$$

$$2\sqrt{\frac{t}{\pi}} \Leftrightarrow \frac{1}{p\sqrt{p}}$$

$$\frac{e^{-at}}{\sqrt{\pi t}} \Leftrightarrow \frac{1}{\sqrt{p+a}}$$

$$\frac{1}{\sqrt{a}}\mathrm{erf}(\sqrt{at}) \Leftrightarrow \frac{1}{p\sqrt{p+a}}$$

$$\frac{e^{at}}{\sqrt{a}}\mathrm{erf}(\sqrt{at}) \Leftrightarrow \frac{1}{(p-a)\sqrt{p}}$$

5. Functions of a Complex Variable

Complex variables introduce a new variable, the complex variable z, which incorporates, as components of z, the x and y coordinates of the xy plane. The complex variable z has the form $z = x+iy$, which can be manipulated the same way as conventional numbers. It is important to know that the symbol i is just another number where $i^2 = -1$.

This is about the application of the methods of the differential and integral calculus to complex numbers and functions of complex variables. The results produce tremendous analytic methods.

5.1 Complex Numbers

The words complex and imaginary are potentially misleading, because complex numbers are not complicated and imaginary operators are not part of someone's imagination. Both words are labels: they are technical terms used to designate a class of numbers. A complex number z is represented by an ordered pair of real numbers x and y written as (x, y).

Multiplication by −1 and √−1 A number can be represented as a distance on a number line. We define steps to the right as positive so that distance AB=+4. Multiply +4 by −1 to get −4 that is the distance AC. Multiply AC by −1 to get back to AB. Clearly multiplication by −1 in effect *rotates* AB and AC by 180°.

```
-8 -7 -6 -5 -4 -3 -2 -1  0  1  2  3  4  5  6  7  8
            C              A              B
```

If +4 is multiplied by √−1 the result is 4√−1. Multiply 4√−1 by √−1 to get −4. Hence multiplication by √−1 two times rotates AB by 180°. And so multiplication by √−1 implements a 90° rotation of AB.

The world has agreed that numbers such as 4√−1 are *imaginary* numbers. To save writing √−1 is replaced by *i* in the mathematical literature.

The ordered pair (x_1, y_1) is a point in the (x, iy) plane that can be reached by starting from the origin, marching along the x-axis for a distance x_1, rotating π/2 radians, and marching parallel to the iy-axis for distance y_1 (Figure 500a).

5 Functions of a Complex Variable

Working with ordered pairs (x, y) does not have much appeal, which is why the world adopted the well known alternative z=x+iy that is easier to work with.

In other words: taking our clue from the rotation operation we use i as a π/2 rotation operator. Then we say iy_1 is a vector we add to vector x_1 so that $z_1 = x_1+iy_1$. This replaces the ordered pair (x_1, y_1). We say z is a complex number whose real part is x and whose imaginary part is y. Keep in mind that x and y are real numbers.

Figure 500 Complex numbers in Cartesian and polar coordinates

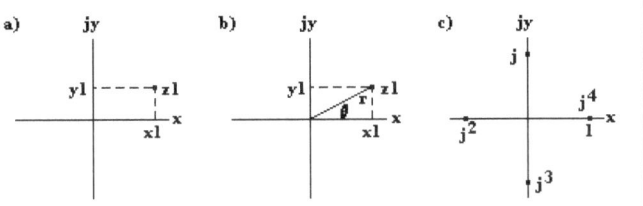

Polar coordinates: If r is the distance from the origin to the point z, then x = r cos θ, and y = r sin θ (Figure 500b). See Euler relation below.

(1) $\quad z = x+iy = r\cos\theta + ir\sin\theta = re^{i\theta}$

(2) $\quad \tan\theta = \dfrac{y}{x}$ so that $\theta = \tan^{-1}\dfrac{y}{x}$

Multiples of i Representing i as a π/2 rotation yields the same results as the √−1 representation (Figure 500c).

(3) $\quad i = e^{i\frac{\pi}{2}} = \cos\dfrac{\pi}{2} + i\sin\dfrac{\pi}{2} = 0 + i1 = i$

(4) $\quad i^2 = e^{i\frac{\pi}{2}2} = e^{i\pi} = \cos\pi + i\sin\pi = -1 + i0 = -1$

(5) $\quad i^3 = e^{i\frac{\pi}{2}3} = e^{i\frac{3\pi}{2}} = \cos\dfrac{3\pi}{2} + i\sin\dfrac{3\pi}{2} = -0 - i1 = -i$

(6) $\quad i^4 = e^{i\frac{\pi}{2}4} = e^{i2\pi} = \cos 2\pi + i\sin 2\pi = 1 + i0 = 1$

Addition The sum of complex numbers is found by adding the two x's and then the y's.

$\quad z_1 + z_2 = (x_1 + iy_1) + (x_2 + iy_2)$

(7) $\quad z_1 + z_2 = (x_1 + x_2) + i(y_1 + y_2)$

Mathematics beyond the Calculus

Multiplication To find the product $z_1 z_2$ multiply z_1 and z_2 in the usual way, while treating i as just another number. Then substitute -1 for i^2.

(8) $\quad z_1 z_2 = (x_1 + iy_1)(x_2 + iy_2)$

$\qquad = x_1 x_2 + x_1 i y_2 + i y_1 x_2 + i y_1 i y_2$

$\qquad = x_1 x_2 + i y_1 i y_2 + i y_2 x_1 + i y_1 x_2$

$\qquad = x_1 x_2 + i^2 y_1 y_2 + i(x_2 y_1 + x_1 y_2)$

$\qquad = (x_1 x_2 - y_1 y_2) + i(x_2 y_1 + x_1 y_2)$

Subtraction Subtraction is defined as addition of positive and negative complex numbers.

(9) $\quad z_1 - z_2 = z_1 + [-z_2] = (x_1 + iy_1) + (-x_2 - iy_2)$

$\qquad = (x_1 - x_2) + i(y_1 - y_2)$

Division Division is facilitated by the complex conjugate concept, where i is replaced by $-i$.

$\qquad$ If $z = x + iy$, then $\bar{z} = x - iy$

$\qquad z\bar{z} = (x + iy)(x - iy) = x^2 - i^2 y^2 + ixy - iyx$

(10) $\quad z\bar{z} = x^2 + y^2 = r^2 = |z|^2 = |z| \times |z|$

$\qquad \dfrac{z_1}{z_2} = \dfrac{z_1}{z_2} \times \dfrac{\bar{z}_2}{\bar{z}_2} = \dfrac{(x_1 + iy_1)(x_2 - iy_2)}{r_2^2} = \dfrac{x_1 x_2 - i^2 y_1 y_2 - i x_1 y_2 + i y_1 x_2}{r_2^2}$

(11) $\quad \dfrac{z_1}{z_2} = \dfrac{x_1 x_2 + y_1 y_2}{r_2^2} + i \dfrac{x_2 y_1 - x_1 y_2}{r_2^2}$

Euler Relation (Figure 500b)

If $r = 1$ then $z = \cos\theta + i\sin\theta$

$\dfrac{dz}{d\theta} = -\sin\theta + i\cos\theta = i(\cos\theta + i\sin\theta) = iz$

(12) $\quad \dfrac{dz}{z} = id\theta$

Integrating $\quad \ln z = i\theta + constant$

If $\theta = 0$ then $z = 1$ so that $\ln 1 = i0 + constant$

However, $\ln 1 = 0$ so that $constant = 0$

$\therefore \ln z = i\theta \implies z = e^{i\theta}$

(13) $\quad e^{i\theta} = \cos\theta + i\sin\theta$

De Moivre's Theorem

(14) $\quad z^n = (re^{i\theta})^n = r^n e^{in\theta} = r^n(\cos n\theta + i\sin n\theta)$

5 Functions of a Complex Variable

Problem 501
show that $(3+2i)-(4-i) = -1+3i$ show that $(2-3i)(-2+i) = -1+8i$

Problem 502
show that $(1-i)^4 = -4$ show that $z^2 - 2z + 2 = (z-1+i)(z-1-i)$

Problem 503
show that $\overline{z_1 - z_2} = \overline{z_1} - \overline{z_2}$ show that $\overline{z_1 z_2} = \overline{z_1}\,\overline{z_2}$ show that $\overline{\left(\dfrac{z_1}{z_2}\right)} = \dfrac{\overline{z_1}}{\overline{z_2}}$

Problem 504
show that $\overline{iz} = -i\,\overline{z}$ show that $\overline{z^n} = (\overline{z})^n$

Problem 505
if $z = x + iy$ show that
a) $z + \overline{z} = 2Re(z)$ b) $z - \overline{z} = 2i\,Im(z)$ c) $|Re(z)| \le |z|$ d) $|Im(z)| \le |z|$

Problem 506
show that $z = 1 + i\sqrt{3} = 2e^{i\pi/3}$ show that $i(1-i\sqrt{3})(\sqrt{3}+i) = 2(1+i\sqrt{3})$

Problem 507
show that $(-2-i2)^7 = 2^{10}(-1+i)$ show that $(\sqrt{3}+i)^{-3} = -i\frac{1}{8}$

Problem 508
show that the square roots of i are $\dfrac{1}{\sqrt{2}}(1+i)$ and $-\dfrac{1}{\sqrt{2}}(1+i)$

Problem 509
show that the cube roots of 1 are 1 $\dfrac{1}{2}(-1+i\sqrt{3})$ $\dfrac{1}{2}(-1-i\sqrt{3})$

Problem 510
if $\omega = \dfrac{1}{2}(-1+i\sqrt{3})$ show that $1 + \omega + \omega^2 = 0$

Problem 511
if k is real show that $(z_1 z_2)^k = z_1^k z_2^k$ $\left(\dfrac{z_1}{z_2}\right)^k = \dfrac{z_1^k}{z_2^k}$

47

5.2 Analytic Functions

Analytic functions are the base on which the subject of functions of complex variables is constructed.

> If function f(z) is defined and *differentiable* at every point in its domain, then f(z) is *analytic*.

A function analytic on the whole complex z plane is referred to as *entire*. If f(z) is differentiable at all points of some neighborhood $|z - z_0| < r$, then f(z) is said to be *analytic* at z_0. If f(z) is analytic at each point of a domain D, then f(z) is *analytic in D*. Since analytic functions are differentiable, they are continuous. Differentiation of sums, products, and quotients of complex functions follow the same rules as for real functions. The essence of an analytic function is that the value of a derivative of f(z) at point z_0 is independent of how z approaches z_0 when forming the derivative. Compare this to real variable derivatives at (x_0, y_0) where derivatives are formed as x approaches x_0 or as y approaches y_0.

The Cauchy-Riemann Equations This famous pair of partial differential equations connects the real and imaginary parts of an analytic function.

(15a) $f(z) = u(x,y) + iv(x,y) \rightarrow \dfrac{\partial u}{\partial x} = \dfrac{\partial v}{\partial y}$ and $\dfrac{\partial u}{\partial y} = -\dfrac{\partial v}{\partial x}$

(15b) $z = re^{i\theta}$ in polar coordinates $\rightarrow r\dfrac{\partial u}{\partial r} = \dfrac{\partial v}{\partial \theta}$ and $\dfrac{\partial u}{\partial \theta} = -r\dfrac{\partial v}{\partial r}$

It follows from the *Cauchy-Riemann Equations* that

(17a) $\dfrac{\partial^2 u}{\partial x^2} = \dfrac{\partial}{\partial x}\dfrac{\partial u}{\partial x} = \dfrac{\partial}{\partial x}\dfrac{\partial v}{\partial y} = \dfrac{\partial^2 v}{\partial y \partial x}$ (17b) $\dfrac{\partial^2 u}{\partial y^2} = \dfrac{\partial}{\partial y}\dfrac{\partial u}{\partial y} = -\dfrac{\partial}{\partial y}\dfrac{\partial v}{\partial x} = -\dfrac{\partial^2 v}{\partial y \partial x}$

$\rightarrow \dfrac{\partial^2 u}{\partial x^2} + \dfrac{\partial^2 u}{\partial y^2} = 0$ (17c) *and by a similar process* $\dfrac{\partial^2 v}{\partial x^2} + \dfrac{\partial^2 v}{\partial y^2} = 0$

Example: show that ln z is analytic and calculate the derivative of ln z.

(16a) $f(z) = \ln z = \ln|z| + i \arg z = \dfrac{1}{2}\ln(x^2 + y^2) + i\arctan\dfrac{y}{x}$

(16b) $\dfrac{\partial u}{\partial x} = \dfrac{x}{x^2 + y^2}$ $\dfrac{\partial u}{\partial y} = \dfrac{y}{x^2 + y^2}$ $\dfrac{\partial v}{\partial x} = \dfrac{-y}{x^2 + y^2}$ $\dfrac{\partial v}{\partial y} = \dfrac{x}{x^2 + y^2}$

(16c) *therefore* $\dfrac{\partial u}{\partial x} = \dfrac{\partial v}{\partial y}$ *and* $\dfrac{\partial u}{\partial y} = -\dfrac{\partial v}{\partial x}$ *hence* ln z *is analytic*

(16d) $\dfrac{d \ln z}{dz} = \dfrac{\partial u}{\partial x} + i\dfrac{\partial v}{\partial x} = \dfrac{x}{x^2 + y^2} + i\dfrac{-y}{x^2 + y^2} = \dfrac{x - iy}{x^2 + y^2} = \dfrac{\bar{z}}{|z|^2} = \dfrac{\bar{z}}{z\bar{z}} = \dfrac{1}{z}$

5 Functions of a Complex Variable

Hence functions u and v satisfy the same differential equation known as *Laplace's Differential Equation*. Functions u and v that satisfy Laplace's Equation are referred to as *harmonic* functions.

Problem 512
form the 2 Cauchy Riemann equations -
show where the equations are valid i.e. where the functions are analytic

a) $w = 2z^2 - z + 1$ everywhere b) $w = az + b$ (a,b complex) everywhere

c) $w = \dfrac{1}{z}$ all z except 0 d) $w = 2x + ixy^2$ nowhere

Problem 513
Let $w = z^3 - 2z = u + iv$ show that u and v satisfy the CR equations

Problem 514
show that $h(z) = \bar{z}$ is NOT analytic i.e. does not satisfy the CR equations

Problem 515
Let $z = x + iy$ and $f = u + iv$, given u find v show that

a) if $u = 2xy$ then $v = -x^2 + y^2 + c$ b) if $u = e^x \cos y$ then $v = e^x \sin y + c$

c) if $u = (e^x + e^{-x})\cos y$ then $v = (e^x - e^{-x})\sin y + c$

Problem 516
Let $w = u + iv = \sqrt{z} = r^{0.5}[\cos(\theta/2) + I\sin(\theta/2)]$ $(0 < \arg z < \pi)$
Show that $u = \sqrt{r + (x/2)}$ $v = \sqrt{r - (x/2)}$

Problem 517
Let $w = u + iv$ and $z = r(\cos\theta + i\sin\theta)$

Show that $\dfrac{\partial u}{\partial r} = \dfrac{1}{r}\dfrac{\partial v}{\partial \theta}$ and $\dfrac{\partial v}{\partial r} = -\dfrac{1}{r}\dfrac{\partial u}{\partial \theta}$

Problem 518 Show where the functions are analytic.
a) $f(z) = xy + iy$ nowhere b) $f(z) = \sin x \cosh y + i \cos x \sinh y$ all z

Problem 519 Show that the functions are harmonic and compute v.

a) $u = \dfrac{1}{2}\ln(x^2 + y^2)$ show that $v = \arctan\dfrac{y}{x} + c$

b) $u = \cos x \cosh y$ show that $v = -\sin x \sinh y + c$

Mathematics beyond the Calculus

5.3 Integration

Closed contours A contour C is a continuous curve in the plane of a complex variable such as z. The starting and ending points are called the *initial* and *terminal* points. If C does not cross itself it is a *simple* curve. Contour C is *closed* if the initial and terminal points coincide. The positive direction of travel on a contour is counterclockwise.

Simple closed contours are the focus of complex variable theory and practice.

Examples of closed contours
The process of computing definite integrals by means of the Residue Theorem (page 52) requires the use of contours three of which are parametrized below.

$z = x \qquad -R \leq x \leq R$ 　　　**Figure 501**

$z = Re^{i\theta} \qquad 0 \leq \theta \leq \pi$

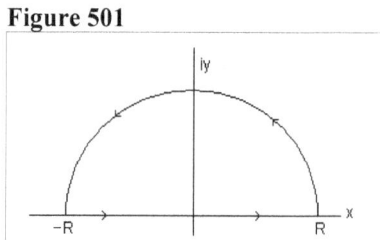

$z = Re^{i\theta} \qquad 0 \leq \theta \leq \pi$ 　　　**Figure 502**

$z = x \qquad -R \geq x \geq -\varepsilon$

$z = \varepsilon e^{i\theta} \qquad \pi \geq \theta \geq 0$

$z = x \qquad \varepsilon \leq x \leq R$

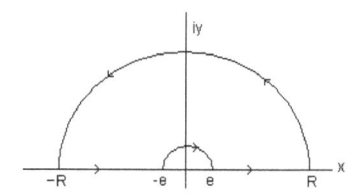

$z = Re^{i\theta} \qquad 0 + d \leq \theta \leq 2\pi - d$ 　　　**Figure 503**

$z = xe^{i(2\pi - d)} \qquad R \geq x \geq \varepsilon$

$z = \varepsilon e^{i\theta} \qquad 2\pi - d \geq \theta \geq d$

$z = xe^{i(0+d)} \qquad \varepsilon \leq x \leq R$

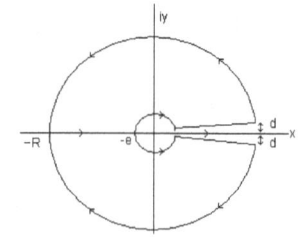

5 Functions of a Complex Variable

Line Integrals The definite integral of a complex function between points a and b is defined by the values of f(z) along some path from a to b.

(18) $\int_a^b f(z)dz = \int_C f(z)dz = \int_C (u+iv)(dx+idy) = \int_C (udx-vdy) + i\int_C (vdx+udy)$

Properties of line integrals

(19a) $\int_a^b f(z)dz = -\int_b^a f(z)dz$

(19b) $\int_a^b k \times f(z)dz = k \times \int_a^b f(z)dz$

(19c) $\int_a^b [f(z)+g(z)]dz = \int_a^b f(z)dz + \int_a^b g(z)dz$

(19d) $\int_a^c f(z)dz = \int_a^b f(z)dz + \int_b^c f(z)dz$

Examples

(20) find the value of $\int_0^{2+i} z^2 dz$ along the straight line path from 0 to $2+i$.

$\int_0^{2+i} z^2 dz = \int_{0,0}^{2,1} [(x^2-y^2)dx - 2xy\,dy] + i\int_{0,0}^{2,1} [2xy\,dx + (x^2-y^2)dy]$

since $x = 2y$ on the path $dx = 2dy$

then $\int_0^{2+i} z^2 dz = \int_0^1 [(6y^2-4y^2)dy] + i\int_0^1 [(8y^2+3y^2)dy] = \frac{2}{3} + i\frac{11}{3}$

(21) find the value of $\int_{-1}^1 \bar{z}\,dz$ along the circular path from π to 0 in the upper

half plane $I_{upper} = \int_{-1}^1 \bar{z}\,dz = \int_\pi^0 e^{-i\theta} i e^{i\theta} d\theta = i\int_\pi^0 d\theta = -i\pi$ ($z = e^{i\theta}$)

and the lower half plane $I_{lower} = \int_{-1}^1 \bar{z}\,dz = \int_\pi^{2\pi} e^{-i\theta} i e^{i\theta} d\theta = i\int_\pi^{2\pi} d\theta = i\pi$

then around the circle → $I_{upper} - I_{lower} = 2\pi i$

Problem 520

Show that $\int_2^{3+i} |z|^2 dz = \frac{20}{3}(1+i)$ over the straight line from 2 to $(3+i)$

Problem 521

Show that $\int_\pi^1 e^z dz = 1+e$ over the straight line path from $i\pi$ to 1

51

Mathematics beyond the Calculus

Cauchy's Theorem is the *key player* of complex variables.
(22) If $f(z)$ is analytic on all points within and on the closed counter C
then $\int_C f(z)dz = 0$

Here is a "proof" of Cauchy's Theorem based on Green's Theorem.

(23) If $f(z)$ is analytic on all points in domain D and on the closed counter C
then $\int_C f(z)dz = i \iint_D \left\{ \frac{\partial f}{\partial x} + i \frac{\partial f}{\partial y} \right\} dxdy$ since $f(z)$ is analytic $f = u + iv$
and u and v satisfy the cauchy - Riemann equations $u_x = v_y$, $u_y = -v_x$

$$\left[\frac{\partial f}{\partial x} \right] + i \left[\frac{\partial f}{\partial y} \right] = \left[\frac{\partial u}{\partial x} + i \frac{\partial v}{\partial x} \right] + i \left[\frac{\partial u}{\partial y} + i \frac{\partial v}{\partial y} \right] = \frac{\partial u}{\partial x} + i \frac{\partial v}{\partial x} + i \frac{\partial u}{\partial y} - \frac{\partial v}{\partial y} = 0$$

Here is a simplistic "proof" of Cauchy's theorem.

(24) If $f(z)$ is analytic on all points within and on the closed counter C
then on any two paths $p1$ and $p2$ from z_1 to z_2 $\int_{p1} f(z)dz = \int_{p2} f(z)dz$
Then on a closed path C the direction of path $p2$ changes and
so does its sign $\rightarrow$ $\int_C dz = \int_{p1} f(z)dz - \int_{p2} f(z)dz = 0$

Examples One way to calculate integrals uses a trick that equates z to functions we can differentiate.

(25) find $\int_C dz$ if C is the unit circle then $z = \cos\theta + i\sin\theta$ and
$dz = (-\sin\theta + i\cos\theta)d\theta \rightarrow \int_0^{2\pi}(-\sin\theta + i\cos\theta)d\theta = 0 + i0 = 0$

(26) find $\int_C zdz$ if C is the unit circle then $z = \cos\theta + i\sin\theta$ and
$dz = (-\sin\theta + i\cos\theta)d\theta \rightarrow \int_0^{2\pi}(\cos\theta + i\sin\theta)(-\sin\theta + i\cos\theta)d\theta =$
$\int_0^{2\pi}[(-2\cos\theta\sin\theta) + i(-\sin^2\theta + \cos^2\theta)]d\theta = -2\times 0 + i(-\pi + \pi) = 0 + i0 = 0$

5 Functions of a Complex Variable

Problem 522 Show that the integrals equal 0.

(a) $\int_C \dfrac{z}{z-2}\,dz$ (b) $\int_C \dfrac{1}{z^2+2z+2}\,dz$ (c) $\int_C z e^{2z}\,dz$ (d) $\int_C \dfrac{1}{\cos z}\,dz$

For any path
Problem 523

show that $\int_{-2}^{-2+i}(z+2)^2\,dz = -i\dfrac{1}{3}$

Problem 524 For any path

show that $\int_0^{\pi+2i} \cos\dfrac{z}{2}\,dz = e + \dfrac{1}{e}$

Problem 525 For any path

show that $\int_{-1}^{i}(1+i4z^3)\,dz = 1+i$

Problem 526 For a path in the right half plane

show that $\int_{-2i}^{2i} \dfrac{dz}{z} = \pi i \quad x>0$

Problem 527 For a path in the left half plane

show that $\int_{-2i}^{2i} \dfrac{dz}{z} = -\pi i \quad x<0$

Problem 528

show that if C is the circle $|z|=2$ and $f(z_0) = \int_C \dfrac{2z^2-z+1}{z-z_0}\,dz$ then $f(1) = 4\pi i$

Problem 529

show that if C is $|z|=2$ then $\int_C \dfrac{dz}{z^4+1} = 0$

Problem 530

show that if C is $|z|=2$ then $\int_C \dfrac{(\cosh z + z^2)}{z(z^2+1)}\,dz = 2\pi i(2-\cos 1)$

53

Mathematics beyond the Calculus

Cauchy's Integral Formula

(27) If $f(z)$ is analytic on all points within and on the closed counter C and if z_0 is a point within C then $f(z_0) = \dfrac{1}{2\pi i} \displaystyle\int_C \dfrac{f(z)}{z-z_0} dz$

Now we derive the formula.

(28) If $f(z)$ is analytic on all points within and on the closed counter C then

$$g(z) = \dfrac{f(z)}{z-z_0}$$

is analytic everywhere in region K

Figure 504

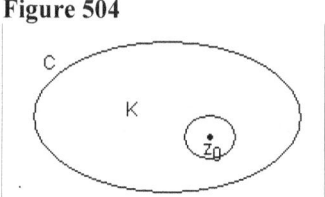

(29) Cauchy's theorem states that $\displaystyle\int_C g(z)dz - \int_K g(z)dz = 0$

on the circle K $z = z_0 + re^{i\theta}$ and $dz = ire^{i\theta} d\theta$ so that

$$\int_K g(z)dz = i\int_0^{2\pi} f(z_0 + re^{i\theta})d\theta = 2\pi i[f(z_0)+r] = 2\pi i f(z_0) \text{ as } r \to 0$$

then $\displaystyle\int_C g(z)dz = 2\pi i f(z_0)$ $\to$ $\dfrac{1}{2\pi i}\displaystyle\int_C g(z)dz = f(z_0)$

replacing $g(z)$ we get Cauchy's formula $\dfrac{1}{2\pi i}\displaystyle\int_C \dfrac{f(z)}{z-z_0} dz = f(z_0)$

and with a change of variable $f(z) = \dfrac{1}{2\pi i}\displaystyle\int_C \dfrac{f(t)}{t-z} dt$

Note – $f(z_0)$ is the residue – see page 57

Cauchy's Formula applied to derivatives

(30) If $f(z)$ is analytic on all points within and on the closed counter C and if z_0 is a point within C then the nth derivative is

$$f^n(z_0) = \dfrac{n!}{2\pi i}\int_C \dfrac{f(z)}{(z-z_0)^{n+1}} dz$$

5 Functions of a Complex Variable

Singularities

A *singularity* (a *singular point*) z_0 of a function f(z) is a point where f(z) is not analytic. And, any open disk about z_0, $|z - z_0| < R$, contains some point z at which f(z) is analytic (Figure 505). Then z_0 is an *isolated singularity* (*isolated singular point*) if f(z) is not analytic at z_0 but is analytic everywhere else in the disk. For example

(31) $f(z) = \dfrac{1}{(z-1)(z+2)}$

has isolated singularities at z = 1, z = –2. which are also known as ***poles***.

Laurent Power Series

Expansion of a function into a power series is based on

(32) $\dfrac{1}{1-\alpha} = 1 + \alpha + \alpha^2 + \alpha^3 + \ldots \qquad |\alpha| < 1$

When a closed contour has singularities inside it a power series expansion the expansion is referred to as a Laurent series.

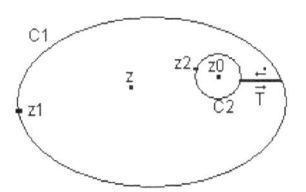

Figure 505

If z_0 is an isolated singularity of *f(z)*, then

(32) $\dfrac{z - z_0}{z_1 - z_0} < 1$ and $\dfrac{z_2 - z_0}{z - z_0} < 1$

and since the two T path integrals cancel (Figure 505)

(33) $\int_C f(z)dz = \int_{C1} f(z)dz + \int_{C2} f(z)dz$

Invoking Cauchy's integral formula -

(34) $\int_C f(z)dz = \dfrac{1}{2\pi i} \int_{C1} \dfrac{f(z_1)}{z_1 - z} dz_1 - \dfrac{1}{2\pi i} \int_{C2} \dfrac{f(z_2)}{z_2 - z} dz_2$

Expanding $1/(z_1 - z)$

(35a) $\dfrac{1}{z_1 - z} = \dfrac{1}{(z_1 - z_0) - (z - z_0)} = \dfrac{1}{(z_1 - z_0)} \times \dfrac{1}{1 - \dfrac{(z - z_0)}{(z_1 - z_0)}}$

(35b) $\dfrac{1}{z_1 - z} = \dfrac{1}{(z_1 - z_0)} + (z - z_0)\dfrac{1}{(z_1 - z_0)^2} + (z - z_0)^2 \dfrac{1}{(z_1 - z_0)^3} + \ldots$

Mathematics beyond the Calculus

Forming Cauchy's integral formula on each term -

(36a) $\dfrac{1}{2\pi i} \int_{C_1} \dfrac{f(z_1)}{z_1 - z} dz_1 = \dfrac{1}{2\pi i} \int_{C_1} \dfrac{f(z_1)}{z_1 - z_0} dz_1 + (z - z_0) \dfrac{1}{2\pi i} \int_{C_1} \dfrac{f(z_1)}{(z_1 - z_0)^2} dz_1$

$+ (z - z_0)^2 \dfrac{1}{2\pi i} \int_{C_1} \dfrac{f(z_1)}{(z_1 - z_0)^3} dz_1 + \ldots$

(36b) on contour C_1 $f(z) = a_0 + a_1(z - z_0) + a_2(z - z_0)^2 + \ldots$

Expanding $-1/(z_2 - z)$ paying attention to the minus sign

(37a) $-\dfrac{1}{z_2 - z} = \dfrac{1}{(z - z_0) - (z_2 - z_0)} = \dfrac{1}{(z - z_0)} \times \dfrac{1}{1 - \dfrac{(z_2 - z_0)}{(z - z_0)}}$

(37b) $\dfrac{1}{z_2 - z} = \dfrac{1}{(z - z_0)} + \dfrac{1}{(z - z_0)^2}(z_2 - z_0) + \dfrac{1}{(z - z_0)^3}(z_2 - z_0)^2 + \ldots$

Forming Cauchy's integral formula on each term -

(38a) $-\dfrac{1}{2\pi i} \int_{C_2} \dfrac{f(z_2)}{(z_2 - z)^{-1}} dz_2 = \dfrac{1}{(z - z_0)} \dfrac{1}{2\pi i} \int_{C_2} \dfrac{f(z_2)}{z_2 - z_0} dz_2$

$+ \dfrac{1}{(z - z_0)^2} \dfrac{1}{2\pi i} \int_{C_2} \dfrac{f(z_2)}{(z_2 - z_0)^{-1}} dz_2 + \dfrac{1}{(z - z_0)^3} \dfrac{1}{2\pi i} \int_{C_2} \dfrac{f(z_2)}{(z_2 - z_0)^{-2}} dz_2 + \ldots$

(38b) on contour C_2 $f(z) = b_1 \dfrac{1}{(z - z_0)} + b_2 \dfrac{1}{(z - z_0)^2} + b_3 \dfrac{1}{(z - z_0)^3} + \ldots$

Add contour C_1 (36b) to contour C_2 (38b) to get the $f(z)$ Laurent series.

(39) $f(z) = a_0 + a_1(z - z_0) + a_2(z - z_0)^2 + \ldots$

$+ b_1 \dfrac{1}{(z - z_0)} + b_2 \dfrac{1}{(z - z_0)^2} + b_3 \dfrac{1}{(z - z_0)^3} + \ldots$

Emphasis - Derivatives of f(z)

(30) If $f(z)$ is analytic on all points within and on the closed counter C and if z_0 is a point within C then the nth derivative is

$f^n(z_0) = \dfrac{n!}{2\pi i} \int_C \dfrac{f(z)}{(z - z_0)^{n+1}} dz$

The coefficients a_n and b_n are derivatives of f(z) except for the n!.

(40a) $a_n = \dfrac{1}{2\pi i} \int_{C_1} \dfrac{f(z_1)}{(z_1 - z_0)^{n+1}} dz_1$ $n = 0, 1, 2, etc$

(40b) $b_n = \dfrac{1}{2\pi i} \int_{C_2} \dfrac{f(z_1)}{(z_1 - z_0)^{-n+1}} dz_1$ $n = 1, 2, 3, etc$

5 Functions of a Complex Variable

Residues
Cauchy's theorem informs us the integral of an analytic function f(z) in a region is zero (equation 22). If a function has a pole at $z = z_0$ a small region around that pole can be carved out (Figure 504 page 54) so that f(z) is analytic in the remaining region. Then a power series expansion in the remaining region is a Laurent expansion (equation 39). The a_j and b_k terms are Cauchy integrals (equations 36 and 38).

Examples important to the theory of residues.

(41a) find $\int_C \frac{1}{z} dz$ if C is the unit circle then $z = \cos\theta + i\sin\theta$ and

$$dz = (-\sin\theta + i\cos\theta)d\theta \to \int_0^{2\pi} \frac{1}{\cos\theta + i\sin\theta}(i^2\sin\theta + i\cos\theta)d\theta = 2\pi i$$

Another way

(41b) find $\int_C \frac{1}{z} dz$ let $z = re^{i\theta}$ $dz = ire^{i\theta} d\theta \to \int_0^{2\pi} \frac{1}{re^{i\theta}} ire^{i\theta} d\theta = 2\pi i$

(42) find $\int_C (z-z_0)^m dz$ if C is a circle of radius r then $z = r(\cos\theta + i\sin\theta)$

and $dz = r(-\sin\theta + i\cos\theta)d\theta = ir(\cos\theta + i\sin\theta)$

$$\to \int_0^{2\pi} r^n(\cos\theta + i\sin\theta)^n ir(\cos\theta + i\sin\theta)d\theta$$

$$= \int_0^{2\pi} ir^{n+1}(\cos\theta + i\sin\theta)^{n+1} d\theta$$

$$= ir^{n+1} \int_0^{2\pi} [\cos(n+1)\theta + i\sin(n+1)\theta]d\theta$$

$$= 2\pi ir^{-1+1} = 2\pi i \text{ when } n = -1, = 0 \text{ for any other integer}$$

$$\therefore \int_C \frac{1}{(z-z_0)} dz = 2\pi i \quad \text{and} \quad \int_C \frac{1}{(z-z_0)^m} dz = 0 \quad \text{when } m \neq 1$$

Poles of Order 1 If f(z) has a pole of order 1 at z_0 then

(43) $f(z) = a_0 + a_1(z-z_0) + a_2(z-z_0)^2 + ...$

$$+ b_1 \frac{1}{(z-z_0)} + b_2 \frac{1}{(z-z_0)^2} + b_3 \frac{1}{(z-z_0)^3} + ...$$

and, the following results show that coefficient b_1 is the pole of order 1 at z_0 residue.

(42) $\int_C \frac{1}{(z-z_0)} dz = 2\pi i \quad \text{and} \quad \int_C \frac{1}{(z-z_0)^m} dz = 0 \quad \text{when } m \neq 1$

Mathematics beyond the Calculus

Examples

(40) $\int_C \dfrac{\cos z}{z^3} = \int_C \left(\dfrac{1}{z^3} - \dfrac{1}{2!}\dfrac{1}{z^1} + \dfrac{1}{4!}z - \dfrac{1}{6!}z^3 + ... \right) dz = 2\pi i \left(-\dfrac{1}{2!} \right)$

(41) $\int_C \exp\dfrac{1}{z^2} dz = \int_C \dfrac{1}{z^2}\left(1 + \dfrac{1}{z^2} + \dfrac{1}{2!}\dfrac{1}{z^4} + \dfrac{1}{3!}\dfrac{1}{z^6} + ... \right) dz = 0$ There is no $\dfrac{1}{z}$ term

Higher Order Poles Computing the value of residues of higher order poles.

Theorem Let a single valued function f(z) satisfy the conditions that for positive integer m a value g(z_0) exists such that the function
$g(z) = (z - z_0)^m f(z)$ is analytic at z_0, and $g(z_0) \neq 0$
Then f(z) has a pole of order m at z_0. And its residue at z_0 is

$\dfrac{1}{(m-1)!} \dfrac{d^{m-1}g(z_0)}{dz} = \dfrac{g^{m-1}(z_0)}{(m-1)!}$ the special symbol for a derivative

Example – find the residue of f(z), which has a pole of order 2.

(42a) $f(z) = \dfrac{\ln z}{(z^2 + 1)^2} = \dfrac{\ln z}{(z-i)^2 (z+i)^2} = \dfrac{1}{(z-i)^2} \dfrac{\ln z}{(z+i)^2}$

The residue at the pole z = i equals the value of the derivative of the second factor of (42a).

(42b) $\dfrac{df(z)}{dz} = \ln z \times \dfrac{d}{dz}\dfrac{1}{(z+i)^2} + \dfrac{1}{(z+i)^2} \dfrac{d \ln z}{dz}$

$= \ln z \times \dfrac{-2}{(z+i)^3} + \dfrac{1}{(z+i)^2}\dfrac{1}{z}$

at the pole $z_0 = i$ the residue =

$\ln i \dfrac{-2}{(i+i)^3} + \dfrac{1}{(i+i)^2}\dfrac{1}{i} = i\dfrac{\pi}{2}\dfrac{-2}{-8i} + \dfrac{1}{-4i} = \dfrac{\pi}{8} + \dfrac{1}{4i}$

at the pole $z_0 = -i$ the residue also equals $\dfrac{\pi}{8} + \dfrac{1}{4i}$

5 Functions of a Complex Variable

This brings us to the residue theorem.

Residue Theorem Let C be a closed contour within and on which $f(z)$ is analytic except for a finite number of poles $z_0, z_1, \ldots, z_n$ inside C. Then if $k_0, k_1, \ldots, k_n$ are the residues of the poles $z_0, z_1, z_2, \ldots, z_n$

Figure 506

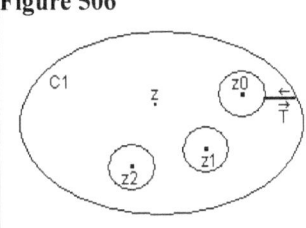

(42) $\quad \int_C f(z)dz = 2\pi i(k_0 + k_1 + \ldots + k_n)$

Problem 531

$\dfrac{z+1}{z^2-2z}$ show that the residues are $-\dfrac{1}{2}\ \dfrac{3}{2}$ and order $m = 1$

Problem 532

tanh z show that the residues are 1 and order $m = 1$

Problem 533

$\dfrac{1-e^{2z}}{z^4}$ show that the residues are $-\dfrac{4}{3}$ and order $m = 3$

Problem 534

$\dfrac{e^{2z}}{(z-1)^2}$ show that the residues are $3e^2$ and order $m = 2$

Problem 535

show that $\int_C \dfrac{1}{z \sin z} dz = 0$ where C is the circle $|z| = 1$

Problem 536

show that $\int_C \dfrac{1}{z^2 \sin z} dz = \dfrac{i\pi}{3}$ where C is the circle $|z| = 1$

Problem 537

show that $\int_C z e^{\frac{1}{z}} dz = i\pi$ where C is the circle $|z| = 1$

59

Mathematics beyond the Calculus

Evaluation of Integrals

Examples using the path of Figure 501 on page 50.

$$\int_0^\infty \frac{1}{x^2+1}\,dx \to \frac{1}{2}\int_{-\infty}^\infty \frac{1}{x^2+1}\,dx$$

$$\to \int_C \frac{1}{z^2+1}\,dz = \int_C \frac{dz}{(z+i)(z-i)} \to \text{residue at } i = 2\pi i \frac{1}{(i+i)} = \pi$$

$$= \lim_{R\to\infty} \int_{-R}^R \frac{e^{ix}}{x^2+1}\,dx + \int_{C_2} \frac{e^{iz}}{z^2+1}\,dz = \pi$$

note that $\int_{C_2}\frac{1}{z^2+1}\,dz \le \int_0^\pi \frac{1}{R^2-1}|dz| = \frac{\pi R}{R^2-1} = 0 \text{ as } R\to\infty$

then $\int_C \frac{1}{z^2+1}\,dz = \lim_{R\to\infty}\int_{-R}^R \frac{1}{x^2+1}\,dx = \pi$ so that $\int_0^\infty \frac{1}{x^2+1}\,dx = \frac{\pi}{2}$

$$\int_0^\infty \frac{\cos x}{x^2+1}\,dx \to \frac{1}{2}\int_{-\infty}^\infty \frac{\cos x}{x^2+1}\,dx \to \frac{1}{2}\text{real}\int_{-\infty}^\infty \frac{e^{ix}}{x^2+1}\,dx$$

$$\to \int_C \frac{e^{iz}}{z^2+1}\,dz = \int_C \frac{e^{iz}\,dz}{(z+i)(z-i)} \to \text{residue at } i = 2\pi i\frac{e^{i\cdot i}}{(i+i)} = \frac{\pi}{e}$$

$$= \lim_{R\to\infty}\int_{-R}^R \frac{e^{ix}}{x^2+1}\,dx + \int_{C_2} \frac{e^{iz}}{z^2+1}\,dz = \frac{\pi}{e}$$

note that $\int_{C_2}\frac{e^{iz}}{z^2+1}\,dz \le \int_0^\pi \frac{1}{R^2-1}|dz| = \frac{\pi R}{R^2-1} = 0 \text{ as } R\to\infty$

then $\int_C \frac{e^{iz}}{z^2+1}\,dz = \lim_{R\to\infty}\int_{-R}^R \frac{e^{ix}}{x^2+1}\,dx = \frac{\pi}{e}$ so that $\int_0^\infty \frac{1}{x^2+1}\,dx = \frac{\pi}{2e}$

Example using the path of the unit circle

on the unit circle $z = e^{i\theta} \to dz = ie^{i\theta}\,d\theta = iz\,d\theta$ and $\sin\theta = \frac{z-z^{-1}}{2i}$

$$I = \int_0^{2\pi} \frac{1}{\frac{5}{4}+\sin\theta}\,d\theta = \int_C \frac{1}{iz\left(\frac{5}{4}+\frac{z-z^{-1}}{2i}\right)}\,dz = \int_C \frac{4}{2z^2+i5z-2}\,dz$$

$$I = \int_C \frac{2}{(z+2i)(z+\frac{1}{2}i)}\,dz = 2\pi i \frac{2}{(-\frac{1}{2}i+2i)} = 2\pi i \frac{2}{(\frac{3}{2}i)} = \frac{8\pi}{3}$$

5 Functions of a Complex Variable

Problem 538

show that $\displaystyle\int_0^\pi \frac{x^2}{(x^2+1)(x^2+4)}dx = \frac{\pi}{6}$

Problem 539

show that $\displaystyle\int_0^\pi \frac{1}{(x^2+4)}dx = \frac{\pi\sqrt{2}}{4}$

Problem 540

show that $\displaystyle\int_0^\infty \frac{x^2}{(x^6+1)}dx = \frac{\pi}{6}$

Problem 541

show that $\displaystyle\int_0^\pi \frac{x^6}{(x^4+1)^2}dx = \frac{3\pi\sqrt{2}}{16}$

Problem 542

show that $\displaystyle\int_0^\infty \frac{\cos bx}{(x^2+1)}dx = \frac{\pi}{2e^b}$ $(b>0)$

Problem 543

show that $\displaystyle\int_{-\infty}^\infty \frac{x}{(x^2+1)(x^2+2x+2)}dx = -\frac{\pi}{5}$

Problem 544

show that $\displaystyle\int_{-\infty}^\infty \frac{\sin x}{(x^2+4x+5)}dx = -\frac{\pi\sin 2}{e}$

Problem 545

show that $\displaystyle\int_0^{2\pi} \frac{1}{5+3\cos\theta}d\theta = \frac{\pi}{2}$

Problem 546

show that $\displaystyle\int_0^\pi \frac{1}{(b+\cos\theta)^2}d\theta = \frac{\pi b}{(b^2-1)^2}$ $(b>1)$

6 Inverse Laplace transform

Inverse transform If we use the Laplace transform to solve a problem in the p domain, then we need an *inverse* transform to return to the t domain.

A return from the *complex frequency domain* p to the t domain is achieved by performing the inverse operation. The operation is known as the Inverse Laplace Transform (equation 1). The inverse transform is an integral on the z plane line from σ–i ∞ to σ+i ∞.

(1) $\quad f(t) = \mathcal{L}^{-1}[F(p)] = \dfrac{1}{2\pi i} \displaystyle\int_{\sigma-i\infty}^{\sigma+i\infty} F(p) e^{tp} \, dp$

Figure 601

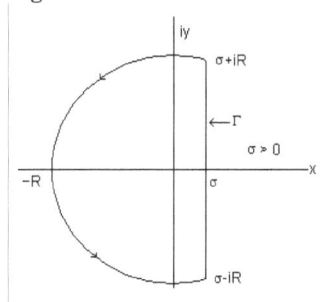

A straightforward method for integrating the inverse transform integral is to convert it to a closed contour Γ in the z plane so that the theory of residues can be used (Figure 601).

(2) $\quad f(t) = \dfrac{1}{2\pi i} \displaystyle\oint_C F(z) e^{tz} \, dz = \sum \text{residues}$

The Γ contour requires all of the poles of F(z) to be in the left hand plane. This is a valid assumption for physical systems.

Example

(3) $\quad F(p) = \dfrac{1}{p(p+\alpha)} = \dfrac{1}{\alpha}\left(\dfrac{1}{p} - \dfrac{1}{p+\alpha}\right) \quad \rightarrow \quad \text{using the trick } f(t) = \dfrac{1}{\alpha}\left(1 - e^{-\alpha t}\right)$

Converting to the z plane we get

(4) $\quad f(t) = \dfrac{1}{2\pi i} \displaystyle\oint_C F(z) e^{tz} \, dz = \dfrac{1}{2\pi i} \displaystyle\oint_C \left(\dfrac{1}{\alpha}\dfrac{1}{z} - \dfrac{1}{\alpha}\dfrac{1}{z+\alpha}\right) e^{tz} \, dz = \sum \text{residues}$

(5) The poles are at z = 0 and z = –α   (< σ)

$\text{residue at } 0 = \lim\limits_{z \to 0} z e^{tz} F(z) = \lim\limits_{z \to 0} \dfrac{e^{tz}}{z+\alpha} = \dfrac{1}{\alpha}$

$\text{residue at } -\alpha = \lim\limits_{z \to -\alpha} (z+\alpha) e^{tz} F(z) = \lim\limits_{z \to -\alpha} \dfrac{e^{tz}}{z} = -\dfrac{e^{-\alpha t}}{\alpha}$

$f(t) = \sum \text{residues} = \dfrac{1}{\alpha}\left(1 - e^{-\alpha t}\right)$

6 Inverse Laplace Transform

Example

(6a) $x(p) = \dfrac{2ap}{(p^2+a^2)^2} \quad \rightarrow \quad x(t) = \dfrac{1}{2\pi i}\int_C e^{zt}\dfrac{2az}{(z^2+a^2)^2}dz$

(6b) $\left[\dfrac{d}{dz}\left(e^{zt}\dfrac{2az}{(z+ia)^2}\right)\right]_{z=ia}$

$= \dfrac{2aia}{(ia+ia)^2}te^{iat} + 2ae^{iat}\left(\dfrac{1}{(ia+ia)^2} + ia\dfrac{-2}{(ia+ia)^3}\right)$

$= \dfrac{2a^2 i}{4i^2 a^2}te^{iat} + 2ae^{iat}\left(\dfrac{1}{-4a^2} + ia\dfrac{-2}{8i^3 a^3}\right)$

$= \dfrac{1}{2i}te^{iat} + e^{iat}\left(\dfrac{1}{-2a} + \dfrac{1}{2a}\right) = \dfrac{1}{2i}te^{iat} + 0$

(6c) residue of pole at $(z = ia) = \dfrac{1}{2i}te^{iat}$

(6d) residue of pole at $(z = -ia) = -\dfrac{1}{2i}te^{-iat}$

(6e) then $x(t) = \dfrac{1}{2i}te^{iat} - \dfrac{1}{2i}te^{-iat} = t\sin at$

Example

(7a) $(D^2 + a^2)^2 x(t) = \cos at \quad\quad D = \dfrac{d}{dt}$ and $x, Dx, D^2 x, D^3 x = 0$ when $t = 0$

(7b) $x(p) = \dfrac{p}{(p^2+a^2)^3} \quad \rightarrow \quad x(t) = \dfrac{1}{2\pi i}\int_C e^{zt}\dfrac{z}{(z^2+a^2)^3}dz$

(7c) $\left[\dfrac{d^2}{dz^2}\left(e^{zt}\dfrac{z}{(z^2+a^2)^3}\right)\right]_{z=ia} = -\dfrac{t}{8a^2}\left(t+\dfrac{i}{a}\right)e^{iat}$

(7d) residue of pole at $(z = ia) = -\dfrac{t}{16a^2}\left(t+\dfrac{i}{a}\right)e^{iat}$

(7e) residue of pole at $(z = -ia) = -\dfrac{t}{16a^2}\left(t-\dfrac{i}{a}\right)e^{-iat}$

(7f) then $x(t) = \dfrac{t}{8a^3}(\sin at - at\cos at)$

Find $f(t)$ by integrating the inverse transform integral by converting it to a closed contour Γ in the z plane and apply the theory of residues.

Problem 601

show that $f(t) = \mathscr{L}^{-1}\left[\dfrac{p+1}{p^2(p-1)}\right] = -2-t+2e^t$

Problem 602

show that $f(t) = \mathscr{L}^{-1}\left[\dfrac{2p^2}{(p^2+1)(p-1)^2}\right] = -\cos t + e^t + te^t$

Problem 603

show that $f(t) = \mathscr{L}^{-1}\left[\dfrac{p}{(p+2)(p-1)(p-3)}\right] = -\dfrac{1}{6}e^t - \dfrac{2}{15}e^{-2t} + \dfrac{3}{10}e^{3t}$

Problem 604

show that $f(t) = \mathscr{L}^{-1}\left[\dfrac{p^2+1}{p(p-1)^2}\right] = -1 + e^t + t^2 e^t$

Problem 605

show that $f(t) = \mathscr{L}^{-1}\left[\dfrac{2p^2+3}{(p+1)^2(p^2+1)^2}\right] = \dfrac{3}{2}e^{-t} + \dfrac{5}{4}te^{-t} - \dfrac{3}{2}\cos t + \dfrac{1}{4}\sin t - \dfrac{1}{4}t\sin t$

Problem 606

show that $f(t) = \mathscr{L}^{-1}\left[\dfrac{p^3}{(p^2+a^2)^3}\right] = \dfrac{1}{8a}\left(3t\sin at + at^2 \cos at\right)$

Problem 607

show that $f(t) = \mathscr{L}^{-1}\left[\dfrac{p^2+a^2}{(p^2-a^2)^2}\right] = t\cosh at$

Problem 608

show that $f(t) = \mathscr{L}^{-1}\left[\dfrac{p}{(p-a)^3}\right] = e^{at}\left(\tfrac{1}{2}at^2 + t\right)$

Problem 609

show that $f(t) = \mathscr{L}^{-1}\left[\dfrac{p}{(p-a)(p-b)}\right] = \dfrac{1}{a-b}\left(ae^{at} - be^{bt}\right) \quad a \neq b$

7 Ordinary Differential Equations

This is about one type of equation – linear differential equations of order n with constant coefficients, which describe many physical phenomena. There are many different types of ordinary differential equations[1], which are solved by various methods. We use two methods.

7.1 Solution by the Laplace Transform

Consider the equation where the function $f(t)$ is the *input*, or *forcing function*, and $y = y(t)$ is the *output* or *response*. (1) $y'''+y''+y = f(t)$

The Laplace transform method is not only straightforward, the method makes initial conditions at $t = 0$ explicit. An important property of the Laplace transform method is that the input function $f(t)$ can be discontinuous. Widder[2] offers a rigorous analysis.

The Laplace transform method for solving ordinary differential equations is implemented by the following steps.
1. Form the Laplace transform of both sides of the differential equation from the t domain to the p domain to produce the *subsidiary equation*, which is an algebraic equation with variable $\mathscr{L}$ (y), Manipulate the *subsidiary equation* to form $\mathscr{L}$ (y) = F(p).
2. Expand $\mathscr{L}$ (y) into partial fractions. Invoke the initial conditions.
3. Perform the inverse transform to produce the solution $y(t) = \mathscr{L}^{-1}$(y).

Techniques for producing the inverse transform in step 3 include partial fraction decomposition, and the translation, derivative, integral, and convolution Laplace transform theorems. These techniques avoid use of integration in the complex plane (Chapter 5). Derivative transforms (2a, 2b, 2c) facilitate the solutions of this type of equation.

(2a) $\mathscr{L}\left[\dfrac{df(t)}{dt}\right] = \mathscr{L}[f'(t)] = pF(p) - f(0)$

(2b) $\mathscr{L}[f''(t)] = p^2 \mathscr{L}[f(t)] - pf(0) - f'(0)$

(2c) $\mathscr{L}[f'''(t)] = p^3 \mathscr{L}[f(t)] - p^2 f(0) - pf'(0) - f''(0)$

---

[1] Tenenbaum & Pollard, *Ordinary Differential Equations*, ISBN 0486 649 407
[2] D. Widder, *Advanced Calculus*, ISBN 0486 661 032

Mathematics beyond the Calculus

Example 1

(3a) $\dfrac{d^2y}{dt^2} - \omega^2 y = 1 \quad \rightarrow \quad$ step 1 $\quad \mathcal{L}[y''] - \omega^2 \mathcal{L}[y] = \mathcal{L}[1]$

(3b) $p^2 \mathcal{L}[y] - py(0) - y'(0) - \omega^2 \mathcal{L}[y] = \dfrac{1}{p}$

(3c) manipulate $(p^2 - \omega^2)\mathcal{L}[y] = py(0) + y'(0) + \dfrac{1}{p}$

(3d) $\mathcal{L}[y] = \dfrac{py(0)}{(p^2 - \omega^2)} + \dfrac{y'(0)}{(p^2 - \omega^2)} + \dfrac{1}{p(p^2 - \omega^2)}$

(3d) step 2 $\mathcal{L}[y] = \dfrac{py(0)}{(p^2 - \omega^2)} + \dfrac{y'(0)}{(p^2 - \omega^2)} - \dfrac{1}{\omega^2 p} + \dfrac{1}{2\omega^2(p+\omega)} + \dfrac{1}{2\omega^2(p-\omega)}$

assume that the initial values are simply numbers - then

(3e) step 3 $y = \mathcal{L}^{-1}[y] = y(0)\cosh \omega t + y'(0)\dfrac{1}{\omega}\sinh \omega t - \dfrac{1}{\omega^2} + \dfrac{e^{-\omega t}}{2\omega^2} + \dfrac{e^{\omega t}}{2\omega^2}$

$y = y(0)\cosh \omega t + y'(0)\dfrac{1}{\omega}\sinh \omega t + \dfrac{1}{\omega^2}(\cosh \omega t - 1)$

Example 2

(4) $y''' + y'' = e^t + t + 1 \quad$ and $\quad y(0) = y'(0) = y''(0) = 0$

$\mathcal{L}[y'''] + \mathcal{L}[y''] = \mathcal{L}[e^t] + \mathcal{L}[t] + \mathcal{L}[1]$

$\{p^3 \mathcal{L}[y] - p^2 y(0) - py'(0) - y''(0)\} + \{p^2 \mathcal{L}[y] - py(0) - y'(0)\} = \dfrac{1}{p-1} + \dfrac{1}{p^2} + \dfrac{1}{p}$

$(p^3 + p^2)\mathcal{L}[y] = \dfrac{1}{p-1} + \dfrac{1}{p^2} + \dfrac{1}{p} \quad \rightarrow \quad \mathcal{L}[y] = \dfrac{2p^2 - 1}{(p^3 + p^2)p^2(p-1)}$

and after many algebraic manipulations and use of a table of transforms

$y = -t + \dfrac{1}{6}t^3 - \dfrac{1}{2}e^{-t} + \dfrac{1}{2}e^t$

Example 3

(5a) $\dfrac{d^2y}{dt^2} - \omega^2 y = 1 \quad \rightarrow \quad \mathcal{L}[y''] - \omega^2 \mathcal{L}[y] = \mathcal{L}[1]$

(5b) $p^2 \mathcal{L}[y] - py(0) - y'(0) - \omega^2 \mathcal{L}[y] = \dfrac{1}{p}$

(5c) $(p^2 - \omega^2)\mathcal{L}[y] = py(0) + y'(0) + \dfrac{1}{p} = \dfrac{p^2 y(0) + py'(0) + 1}{p}$

(5d) $\mathcal{L}[y] = \dfrac{p^2 y(0) + py'(0) + 1}{p(p^2 - \omega^2)} \quad \rightarrow \quad y = \mathcal{L}^{-1}[y]$

7 Ordinary Differential Equations

Example 4
This example uses a trick that deals with the p in *py(0)*.

(6) $\dfrac{py(0)}{2p^2+5p-3} = \dfrac{py(0)}{(2p-1)(p+3)} = \dfrac{y(0)}{2\times 3.5}\left(\dfrac{p}{(p-\frac{1}{2})}+\dfrac{1}{(p+3)}\right)$

the first term does not have a simple inverse transform
the trick is to manipulate the expression as follows

Let $G(p) = \dfrac{py(0)}{2p^2+5p-3} = \dfrac{1}{2}\times\dfrac{2py(0)}{p^2+2\frac{5}{4}p+\frac{25}{16}-\frac{3}{2}-\frac{25}{16}}$

$= \dfrac{py(0)}{(p+\frac{5}{4})^2+(-\frac{3}{2}-\frac{25}{16})} = \dfrac{py(0)}{(p+a)^2-(b^2)}$ where $a = \frac{5}{4}$ $b^2 = \frac{3}{2}+\frac{25}{16}$

$G(p) = y(0)\dfrac{p+a-a}{(p+a)^2-(b^2)} = y(0)\dfrac{(p+a)}{(p+a)^2-(b^2)} - y(0)\dfrac{a}{b}\dfrac{b}{(p+a)^2-(b^2)}$

Solve this equation using the trick.

(7) $(2D^2+5D-3)y = 0 \rightarrow 2\mathcal{L}[y''] + 5\mathcal{L}[y'] - 3\mathcal{L}[y] = 0$

$2\{p^2\mathcal{L}[y] - py(0) - y'(0)\} + 5\{p\mathcal{L}[y] - y(0)\} - 3\mathcal{L}[y] = 0$

$(2p^2+5p-3)\mathcal{L}[y] = 2py(0) + 2y'(0) - 5y(0)$

$F(p) = \mathcal{L}[y] = \dfrac{2py(0)+2y'(0)-5y(0)}{2p^2+5p-3} = \dfrac{1}{2}\dfrac{2py(0)+2y'(0)-5y(0)}{p^2+\frac{5}{2}p-\frac{3}{2}}$

$= \dfrac{1}{2}\dfrac{2py(0)+2y'(0)-5y(0)}{(p+\frac{5}{4})^2+(-\frac{3}{2}-\frac{25}{16})} = \dfrac{1}{2}\dfrac{2py(0)+2y'(0)-5y(0)}{(p+a)^2-(b^2)}$

$= y(0)\dfrac{(p+a)}{(p+a)^2-(b^2)} - y(0)\dfrac{a}{b}\dfrac{b}{(p+a)^2-(b^2)} + \dfrac{b}{b}\times\dfrac{-2.5y(0)+y'(0)}{(p+a)^2-(b^2)}$

$y = \mathcal{L}^{-1}[F(p)] = y(0)[e^{-at}(\cosh bt - \frac{a}{b}\sinh bt] + (-2.5y(0)+y'(0))\frac{1}{b}\sinh bt$

Example 5 The trick +2−2 cancels a (p+2) factor created in the *py(0)* term.

(8) $\dfrac{d^2y}{dx^2} + 4\dfrac{dy}{dx} + 4y = 0 \rightarrow \mathcal{L}[y^{(2)}] + 4\mathcal{L}[y^{(1)}] + 4\mathcal{L}[y] = \mathcal{L}[0] = 0$

$(p^2+4p+4)\mathcal{L}[y] = py(0) + y'(0) + 4y(0)$

$F(p) = \mathcal{L}[y] = \dfrac{py(0)+y'(0)+4y(0)}{(p+2)^2} = \dfrac{(p+2-2)y(0)+y'(0)+4y(0)}{(p+2)^2}$

$= \dfrac{(p+2)y(0)}{(p+2)^2} + \dfrac{y'(0)+2y(0)}{(p+2)^2} = \dfrac{y(0)}{(p+2)} + \dfrac{y'(0)+2y(0)}{(p+2)^2}$

$y = \mathcal{L}^{-1}[F(p)] = y(0)e^{-2x} + [y'(0)+2y(0)]xe^{-2x}$

7.2 Solution by Differential Operator

Differential operators offer another way to solve differential equations of order n with constant coefficients. The solution y of a non *homogeneous* linear differential equation (9) is the sum of a particular integral y_p of the equation and a solution y_c of the corresponding homogeneous equation. Finding the particular integral y_p usually requires special methods, while finding y_c is straightforward.

(9) $\quad a_n \dfrac{d^n y(x)}{dx^n} + a_{n-1} \dfrac{d^{n-1} y(x)}{dx^{n-1}} + \ldots + a_1 \dfrac{dy(x)}{dx} + a_0 y = f(x)$

When $f(x) = 0$ the equation is referred to as homogeneous.

Complementary solution y_c The higher order derivatives in the equation "hint" that the exponential function is a possible solution, because the derivatives of the exponential are exponentials.

(10) $\quad y = e^{mx} \quad \to \quad \dfrac{d^k y}{dx^k} = m^k e^{mx}$

The immediate question is what values of m are solutions?

(11a) $\quad a_n \dfrac{d^n}{dx^n} e^{mx} + a_{n-1} \dfrac{d^{n-1}}{dx^{n-1}} e^{mx} + \ldots + a_1 \dfrac{d}{dx} e^{mx} + a_0 e^{mx} = 0$

(11b) $\quad a_n m^n e^{mx} + a_{n-1} m^{n-1} e^{mx} + \ldots + a_1 m e^{mx} + a_0 e^{mx} = 0$

Since the exponential does not equal zero for any m and x, divide by e^{mx} to create the *characteristic, or auxiliary*, equation.

(12a) $\quad a_n m^n + a_{n-1} m^{n-1} + \ldots + a_1 m + a_0 = 0$

(12a) the polynomial has n roots $m_n, m_{n-1}, m_{n-2}, \ldots, m_1$

The roots create the solutions y_k whose sum y_c is the general solution.

(13a) $\quad y_n = e^{m_n x} \quad y_{n-1} = e^{m_{n-1} x} \quad \ldots \quad y_2 = e^{m_2 x} \quad y_1 = e^{m_1 x}$

(13b) $\quad y_c = c_n e^{m_n x} + c_{n-1} e^{m_{n-1} x} + \ldots + c_2 e^{m_2 x} + c_1 e^{m_1 x}$

> The c constants' values are determined by initial conditions.

The concept of a differential operator simplifies the writing of equations

(14) *define the operators* $D = \dfrac{d}{dx} \quad D^n = \dfrac{d^n}{dx^n} \quad n = 1, 2, 3, \ldots$

7 Ordinary Differential Equations

Exponential Shift Theorem for Polynomial Operators – This theorem is one of many theorems in the calculus of operators.

(15) $D^n[e^{mx} f(x)] = e^{mx}(D+m)^n f(x)$

proof by mathematical induction

$n = 1 \rightarrow D^1[e^{mx} f(x)] = me^{mx} f(x) + e^{mx} Df(x) = e^{mx}(D+m)^1 f(x)$

$n+1 \rightarrow D^{n+1}[e^{mx} f(x)] = D[e^{mx}(D+m)^n f(x)]$

$\qquad = e^{mx} D(D+m)^n f(x) + me^{mx}(D+m)^n f(x)$

$\qquad = e^{mx}(D+m)(D+m)^n f(x)$

$D^{n+1}[e^{mx} f(x)] = e^{mx}(D+m)^{n+1} f(x) \quad$ qed

Example 1 – roots real and distinct

(16) $a_2 \dfrac{d^2 y}{dx^2} + a_1 \dfrac{dy}{dx} + a_0 y = 0 \rightarrow (a_2 D^2 + a_1 D + a_0)y = 0 \rightarrow a_2 m^2 + a_1 m + a_0 = 0$

(17) If the roots of the auxiliary equation are m_1 and m_2, then

$(a_2 D^2 + a_1 D + a_0)y = a_0(D - m_1)(D - m_2)y = 0$

(18a) solutions of $(D - m_1)y = 0$ and $(D - m_2)y = 0$ are solutions of (16)

(18b) $(D - m_n)y = 0 \rightarrow \dfrac{dy}{dx} - m_n y = 0$

$\qquad \rightarrow \dfrac{dy}{dy} = m_n dx \rightarrow \ln y = m_n x + \ln c_n \rightarrow y = c_n e^{m_n x} \quad n = 1, 2$

(19) Therefore $y_c = c_1 e^{m_1 x} + c_2 e^{m_2 x}$

Example 2 – roots real and distinct

(20) $(2D^2 + 5D - 3)y = 0 \rightarrow 2m^2 + 5m - 3 = (m - \tfrac{1}{2})(m + 3) \rightarrow y_c = c_1 e^{-0.5x} + c_2 e^{-3x}$

Example 3 – roots real and multiple

(21) $(D^2 - 4D + 4)y = 0 \rightarrow m^2 - 4m + 4 = (m-2)(m-2) \rightarrow e^{2x}$

then $(D^2 - 4D + 4)e^{2x} f(x) = (D-2)^2 e^{2x} f(x) = e^{2x}(D - 2 + 2)^2 f(x)$

$D^2 f(x) \rightarrow Df(x) = c_2 \rightarrow f(x) = c_1 + c_2 x$

so that $y_c = (c_1 + c_2 x)e^{2x}$ is a solution and in general

$y_c = (c_1 + c_2 x + c_3 x^2 + ... + c_n x^{n-1})e^{ax}$

Example 4 – roots complex and distinct

(22) $(D^2 - 2D + 2)y = 0 \rightarrow m^2 - 2m + 2 = [m - (1+i)][m - (1-i)]$

$y_c = c_1 e^{(1+i)x} + c_2 e^{(1-i)x} = e^x(c_1 e^{ix} + c_2 e^{-ix}) = e^x[(c_1 + c_2)\cos x + i(c_1 - c_2)\sin x]$

69

Mathematics beyond the Calculus

Particular solution y_p

(a) Reduction of order method
Rewrite general equation (9) in differential operator format and then factor it to expose the roots r_k.

(9) $a_n \dfrac{d^n y(x)}{dx^n} + a_{n-1} \dfrac{d^{n-1} y(x)}{dx^{n-1}} + ... + a_1 \dfrac{dy(x)}{dx} + a_0 y = f(x)$

(23a) $a_n D^n + a_{n-1} D^{n-1} + ... + a_1 D^1 + a_0 = f(x)$
(23b) $a_0 (D - r_1)(D - r_2)(D - r_3)....(D - r_{n-1})(D - r_n) y = f(x)$

Start the reduction process.
(24a) Define $(D - r_1) u = f(x)$
(24b) therefore $u = a_0 (D - r_2)(D - r_3)....(D - r_{n-1})(D - r_n) y$

Solve for u. Repeat the process with the next root, etc.
(25a) Define $(D - r_2) v = u$
(25b) therefore $v = a_0 (D - r_3)....(D - r_{n-1})(D - r_n) y$

Example

(26) Solve the equation $(D^2 - 2D^2 + D) y = x$
    factor $(D-1)(D-1) Dy = x$ → first reduction $(D-1) u = x$
    transform x to λ → $\mathscr{L}[Du] - \mathscr{L}[u] = \mathscr{L}[x]$ → $\lambda \mathscr{L}[u] - \mathscr{L}[u] = 1/\lambda^2$

$\mathscr{L}[u] = \dfrac{1}{\lambda^2 (\lambda - 1)} = -\dfrac{1}{\lambda} - \dfrac{1}{\lambda^2} + \dfrac{1}{\lambda - 1}$ → $u = -1 - x + e^x$

consequently $(D-1) Dy = u = -1 - x + e^x$

(27) second reduction → let $v = Dy$ → $(D-1) v = u = -1 - x + e^x$
    transform x to λ → $\mathscr{L}[Dv] - \mathscr{L}[v] = \mathscr{L}[-1 - x + e^x]$

→ $\lambda \mathscr{L}[v] - \mathscr{L}[v] = -\dfrac{1}{\lambda} - \dfrac{1}{\lambda^2} + \dfrac{1}{\lambda - 1}$

$\mathscr{L}[v] = -\dfrac{1}{\lambda(\lambda-1)} - \dfrac{1}{\lambda^2(\lambda-1)} + \dfrac{1}{(\lambda-1)^2}$

$\mathscr{L}[v] = \dfrac{1}{\lambda} - \dfrac{1}{(\lambda-1)} + \dfrac{1}{\lambda} + \dfrac{1}{\lambda^2} - \dfrac{1}{(\lambda-1)} + \dfrac{1}{(\lambda-1)^2} = \dfrac{2}{\lambda} + \dfrac{1}{\lambda^2} - \dfrac{2}{(\lambda-1)} + \dfrac{1}{(\lambda-1)^2}$

$v = 2 + x - 2e^x + xe^x = 2 + x + (x - 2) e^x$
and so $Dy_p = 2 + x + (x - 2) e^x$
$y_p = 2x + \tfrac{1}{2} x^2 + (\tfrac{1}{2} x^2 - 2x) e^x + (x - 2) e^x = 2x + \tfrac{1}{2} x^2 + (\tfrac{1}{2} x^2 - x - 2) e^x$

7 Ordinary Differential Equations

(28) $y = y_c + y_p = (c_1 + xc_2)e^x + c_3 + 2x + \frac{1}{2}x^2 + (\frac{1}{2}x^2 - x - 2)e^x$

$= (c_1 + xc_2 + \frac{1}{2}x^2 - x - 2)e^x + c_3 + 2x + \frac{1}{2}x^2$

$= \{c_1 - 2 + x(1 + c_2) + \frac{1}{2}x^2\}e^x + c_3 + 2x + \frac{1}{2}x^2$

$= \{c_4 + c_5 x + \frac{1}{2}x^2\}e^x + c_3 + 2x + \frac{1}{2}x^2$

(b) Undetermined coefficients method

One of the special methods is the *method of undetermined coefficients*[3]. The method applies to any function f(x) such that f(x) and a finite number of its successive derivatives form a set of linearly dependent functions as shown in (29).

(29) $b_0 f^{(n)}(x) + b_1 f^{(n-1)}(x) + ... + b_{n-1} f^{(1)}(x) + b_n f^{(0)}(x) = 0$

Example of linearly dependent functions.

(30a) $f(x) = x^2 e^{3x} \rightarrow f^{(1)}(x) = (3x^2 + 2x)e^{3x}$
$\rightarrow f^{(2)}(x) = (9x^2 + 12x + 2)e^{3x} \rightarrow f^{(3)}(x) = (27x^2 + 54x + 18)e^{3x}$

(30b) then $f^{(3)}(x) - 9 f^{(2)}(x) + 27 f^{(1)}(x) - 27 f(x) = 0$

Since (29) is a *homogeneous* linear differential equation with constant coefficients f(x) must be expressible as a sum of terms with the following form.

(31) $cx^p e^{qx} \qquad cx^p e^{\alpha x} \cos \beta x \qquad cx^p e^{\alpha x} \sin \beta x$

where c is any constant, $p \geq 0$, and q, α, β are any real constants

If f(x) equals some power of x, the y_p must be some polynomial in x, whose derivatives will satisfy (23).

(23a) $a_n D^n + a_{n-1} D^{n-1} + ... + a_1 D^1 + a_0 = f(x)$

(23b) $a_0 (D - r_1)(D - r_2)(D - r_3)....(D - r_{n-1})(D - r_n) y = f(x)$

Example 1 – a_0 does not equal zero means degree of y_p equals highest power of x in f(x).

(32a) $(D^2 - 1) y_p = x^2 \rightarrow$ let $y_p = Ax^2 + Bx + C$

(32b) $(D^2 - 1)(Ax^2 + Bx + C) = 2A - Ax^2 - Bx - C = x^2$
$-A = 1 \quad B = 0 \quad 2A - C = 0 \rightarrow A = -1, \ B = 0, \ C - 2$

(32c) $y_p = -x^2 - 2$

---

[3] Tenenbaum & Pollard, *Ordinary differential Equations*, ISBN 0486 649 407

Mathematics beyond the Calculus

Example 2 – a_0 equals zero means degree of y_p equals highest power of x in f(x) plus the number of roots r that equal 0.

(33) $g(D)D^r y_p = cx^p \quad \rightarrow \quad$ then let $y_p = Ax^{p+r} + Bx^{p+r-1} + \ldots + Cx^{r+1} + Dx^r$

Example 3 – a_0 equals zero means degree of y_p equals $1+2 = 3$.

(34a) $(D^4 + D^2)y_p = 2x \quad \rightarrow \quad (D^2+1)D^2 y_p = 2x \quad \rightarrow \quad$ let $y_p = x^2(Ax+B)$

(34b) $(D^4 + D^2)x^2(Ax+B) = 6Ax + 2B = 2x \quad \rightarrow \quad A = 1/3, \ B = 0$

(34c) $y_p = \frac{1}{3}x^3$

Example 4

(35) $\dfrac{d^2 y}{dt^2} - \omega^2 y = 1 \quad \rightarrow \quad (D^2 - \omega^2)y = 1 \quad \rightarrow \quad m^2 - \omega^2 = (m+\omega)(m-\omega) = 0$

$y_c = c_1 e^{\omega t} + c_2 e^{-\omega t}$

Let $y_p = A$ (a constant) then evaluate A by assuming $y = A$

$0 - \omega^2 A = 1 \quad \rightarrow \quad A = -\dfrac{1}{\omega^2} = y_p$

and $y = y_c + y_p = c_1 e^{\omega t} + c_2 e^{-\omega t} - \dfrac{1}{\omega^2}$ which one can verify is a solution

Given 2 initial conditions we can solve for c_1 and c_2

Example 5

(24) $y''' + y'' = e^t + t + 1 \quad$ and $\quad y(0) = y'(0) = y''(0) = 0$

$(D^3 + D^2)y = e^t + t + 1 \quad \rightarrow \quad m^3 + m^2 = 0 \quad \rightarrow \quad m^2(m+1) = 0$

$y_c = (c_1 + c_2 t)e^{0t} + c_3 e^{-t}$

$y_p = Ae^t + t^2(Bt + C) \quad \rightarrow \quad y_p''' + y_p'' = 2Ae^t + 6Bt + 6B + 2C = e^t + t + 1$

$A = 1/2 \quad B = 1/6 \quad C = 0$

$y_p = \frac{1}{2}e^t + \frac{1}{6}t^3$

$y = y_c + y_p = c_1 + c_2 t + c_3 e^{-t} + \frac{1}{2}e^t + \frac{1}{6}t^3$

7 Ordinary Differential Equations

Problem 701

Show that the solution to $\dfrac{dy}{dx} - y = e^x$ is $y = [x + y(0)]e^x$

Problem 702

Show that the solution to $\dfrac{dy}{dx} + y = e^{-x}$ is $y = [x + y(0)]e^{-x}$

Problem 703

Show that the solution to $\dfrac{d^2y}{dx^2} - 2\dfrac{dy}{dx} + 5y = 0$ is $y = (2\cos 2x + \sin 2x)e^x$

when $y(0) = 2$ and $y'(0) = 4$

Problem 704

Show that the solution to $\dfrac{d^2y}{dx^2} - 5\dfrac{dy}{dx} - 6y = e^{3x}$ is $y = \dfrac{10}{21}e^{6x} + \dfrac{45}{28}e^{-x} - \dfrac{1}{12}e^{3x}$

when $y(0) = 2$ and $y'(0) = 1$

Problem 705

Show that the solution to $\dfrac{dy(t)}{dt} + y(t) = \cos t$

is $y = \dfrac{1}{2}(\cos t + \sin t - e^{-t}) + y(0)e^{-t}$ hint $(p^2 + 1) = (p+i)(p-i)$

Problem 706

Show that the solution to $\dfrac{d^2y}{dt^2} - y = t$ is $y = -t + 2\sinh t$

when $y(0) = 0$ and $y'(0) = 1$

Problem 707

Show that the solution to $\dfrac{d^2y}{dx^2} - \dfrac{dy}{dx} - 2y = 5\sin x$

is $y = \dfrac{1}{3}e^{2x} + \dfrac{1}{6}e^{-x} - \dfrac{3}{2}\sin x + \dfrac{1}{2}\cos x$ when $y(0) = 1$ and $y'(0) = -1$

Problem 708

Show that the solution to $y'' + y' + y = t^2$ is

$y = t^2 - 2t + e^{-\frac{t}{2}}\left(\cos\dfrac{\sqrt{3}}{2}t + \dfrac{7\sqrt{3}}{3}\sin\dfrac{\sqrt{3}}{2}t\right)$ when $y(0) = 1$ and $y'(0) = 1$

Mathematics beyond the Calculus

8 Systems of Ordinary Differential Equations

A system of ordinary differential equations (ODE) involves two or more ODEs such as these two first order equations where x and y are dependent variables and t is an independent variable.

(1a) $2\dfrac{dx}{dt} - x + \dfrac{dy}{dt} + 4y = 1$ (1b) $\dfrac{dx}{dt} - \dfrac{dy}{dt} = t - 1$

A pair of functions x(t) and y(t) is a solution to this system if this pair satisfies both differential equations.

8.1 Solution by Differential Operator

Write equations (1) in operator notation where D = d/dt (Section 7.2).
(2a) $(2D-1)x + (D+4)y = 1$ (2b) $Dx - Dy = t - 1$

Example 1
The process here is similar to the process of elimination in systems of algebraic equations. Eliminate y by multiplying (2a) by D and (2b) by D+4. Then add the modified equations to get (3c) whose particular solution in (4) is $y_p = x_p(t)$.

(3a) D times (2a) produces $(2D^2 - D)x + D(D+4)y = D1$
(3b) $(D+4)$ times (2b) produces $(D+4)Dx - (D+4)Dy = (D+4)(t-1)$
and their sum is
(3c) $(3D^2 + 3D)x = D1 + D(t-1) + 4(t-1) = 0 + (1-0) + 4(t-1) = 4t - 3$
$(3D^2 + 3D)x = 4t - 3$

(4) Solve the (3c) equation $(D+1)Dx = \tfrac{4}{3}t - 1$

first reduction $(D+1)u = \tfrac{4}{3}t - 1$ and $u = Dx$

transform t to p → $\mathscr{L}[Du] + \mathscr{L}[u] = \mathscr{L}[\tfrac{4}{3}t - 1]$

→ $p\mathscr{L}[u] + \mathscr{L}[u] = \dfrac{4}{3p^2} - \dfrac{1}{p}$

$\mathscr{L}[u] = \dfrac{1}{(p+1)}\left(\dfrac{4}{3p^2} - \dfrac{1}{p}\right) = \dfrac{4}{3}\left(-\dfrac{1}{p} + \dfrac{1}{p^2} + \dfrac{1}{p+1}\right) - \left(\dfrac{1}{p} - \dfrac{1}{p+1}\right)$

$\mathscr{L}[u] = -\dfrac{7}{3p} + \dfrac{4}{3p^2} + \dfrac{7}{3}\dfrac{1}{p+1}$ → $u(t) = -\dfrac{7}{3} + \dfrac{4}{3}t + \dfrac{7}{3}e^{-t}$

consequently $Dx = u = -\dfrac{7}{3} + \dfrac{4}{3}t + \dfrac{7}{3}e^{-t}$ → $x_p = x_p(t) = -\dfrac{7}{3}t + \dfrac{2}{3}t^2 - \dfrac{7}{3}e^{-t}$

8 Systems of Ordinary Differential Equations

Add the complementary solution x_c to x_p in (4) to get x(t) (5a). Substituting x(t) (5a) into (2b) produces y(t).

(5a) *per 7.2 the y_c solution to $3(D+1)Dx = 0$ is $x_c = c_1 + c_2 e^{-t}$*

so that $x(t) = x_c + x_p = c_1 + c_2 e^{-t} - \frac{7}{3}t + \frac{2}{3}t^2$

(5b) *substituting (5a) into (2b)* $D(c_1 + c_2 e^{-t} - \frac{7}{3}t + \frac{2}{3}t^2) - Dy = t - 1$

$(0 - c_2 e^{-t} - \frac{7}{3} + \frac{4}{3}t) - Dy = t - 1$

$Dy = -c_2 e^{-t} - \frac{4}{3} + \frac{1}{3}t \quad \rightarrow \quad y(t) = c_2 e^{-t} - \frac{4}{3}t + \frac{1}{6}t^2 + c_3$

There are 3 constants c_1, c_2, and c_3. But the order of equation (2a) is 2, which means only 2 constants are contained in equations (2). To eliminate one constant substitute x(t) and y(t) in equation (2a), and perform the manipulations to replace c_3.

(6a) $(2D-1)(c_1 + c_2 e^{-t} - \frac{7}{3}t + \frac{2}{3}t^2) + (D+4)(c_2 e^{-t} - \frac{4}{3}t + \frac{1}{6}t^2 + c_3) = 1$

(6b) *perform the manipulations to get*

$-6 - c_1 + 4c_3 = 1 \quad \rightarrow \quad 4c_3 = c_1 + 7 \quad \rightarrow \quad c_3 = \frac{c_1 + 7}{4}$

Equivalent triangular system This is a method that produces the correct number of constants, and sometimes is more efficient, when equations such as (7) are converted to "triangular" form (8).

(7a) $f_1(D)x + g_1(D)y = h_1(t)$ where $f_1(D)$ operates on x, etc
(7b) $f_2(D)x + g_2(D)y = h_2(t)$

The conversion process is simply the process of eliminating g_1 in (7a). The process can be extended to 3 or more simultaneous equations.
(8a) $f_3(D)x \qquad = h_3(t)$
(8b) $f_4(D)x + g_4(D)y = h_4(t)$

Example 2
(9a) $(3D-1)x + 4y = t$ (9b) $Dx - Dy = t - 1$

Eliminate y in (9b) by multiplying (9a) by D/4 and adding it to (9b) to produce (10b).
(10a) $(D/4)(3D-1)x + (D/4)4y + Dx - Dy = (D/4)t + t - 1$

$(D/4)(3D-1)x + Dx = (D/4)t + t - 1$

$(D)(3D-1)x + 4Dx = Dt + 4t - 4 = 4t - 3$

(10b) $(3D^2 + 3D)x = 4t - 3 \quad \rightarrow \quad (D+1)Dx = \frac{4}{3}t - 1 \quad \textit{same as (4)}$

Mathematics beyond the Calculus

8.2 Solution by Laplace Transform

Here we solve (2) by Laplace Transform
(2a) $(2D-1)x+(D+4)y=1$ (2b) $Dx-Dy=t-1$

(11a) $2\dfrac{dx}{dt}-x+\dfrac{dy}{dt}+4y=1$ (11b) $\dfrac{dx}{dt}-\dfrac{dy}{dt}=t-1$

(12a) $2\mathscr{L}[x']-\mathscr{L}[x]+\mathscr{L}[y']+4\mathscr{L}[y]=\mathscr{L}[1]$
$2p\mathscr{L}[x]-2x(0)-\mathscr{L}[x]+p\mathscr{L}[y]-y(0)+4\mathscr{L}[y]=1/p$
$(2p-1)\mathscr{L}[x]+(p+4)\mathscr{L}[y]=\dfrac{1}{p}+2x(0)+y(0)$

(12b) $\mathscr{L}[x']-\mathscr{L}[y']=\mathscr{L}[t-1]$
$p\mathscr{L}[x]-x(0)-p\mathscr{L}[y]+y(0)=\mathscr{L}[t-1]$
$p\mathscr{L}[x]-p\mathscr{L}[y]=\dfrac{1}{p^2}-\dfrac{1}{p}+x(0)-y(0)$

Assume the initial conditions equal zero.

(12a) $(2p-1)\mathscr{L}[x]+(p+4)\mathscr{L}[y]=\dfrac{1}{p}$

(12b) $\mathscr{L}[x]-\mathscr{L}[y]=\dfrac{1}{p^3}-\dfrac{1}{p^2}$

Multiply (12b) by $(2p-1)$, and subtract (12b) from (12a).

(13) $\{(p+4)+(2p-1)\}\mathscr{L}[y]=\dfrac{1}{p}-(2p-1)\left(\dfrac{1}{p^3}-\dfrac{1}{p^2}\right)$

$(3p+3)\mathscr{L}[y]=\dfrac{3}{p}-\dfrac{3}{p^2}+\dfrac{1}{p^3}$ → $\mathscr{L}[y]=\dfrac{1}{p(p+1)}-\dfrac{1}{p^2(p+1)}+\dfrac{1}{3p^3(p+1)}$

Observe that the transform produces values for the constants.

(14) $\mathscr{L}[y]=\dfrac{1}{p}-\dfrac{1}{(p+1)}+\dfrac{1}{p}-\dfrac{1}{p^2}-\dfrac{1}{(p+1)}+\dfrac{1}{3}\left(\dfrac{1}{p}-\dfrac{1}{p^2}+\dfrac{1}{p^3}-\dfrac{1}{(p+1)}\right)$

$\mathscr{L}[y]=\dfrac{2}{p}-\dfrac{2}{(p+1)}-\dfrac{1}{p^2}+\dfrac{1}{3}\left(\dfrac{1}{p}-\dfrac{1}{p^2}+\dfrac{1}{p^3}-\dfrac{1}{(p+1)}\right)$

$\mathscr{L}[y]=\dfrac{7}{3p}-\dfrac{7}{3(p+1)}-\dfrac{4}{3p^2}+\dfrac{1}{3p^3}$ → $y(t)=\dfrac{7}{3}-\dfrac{4}{3}t+\dfrac{1}{6}t^2-\dfrac{7}{3}e^{-t}$

(15) from (11b) $\dfrac{dx}{dt}=\dfrac{dy}{dt}+t-1=0-\dfrac{4}{3}+\dfrac{1}{3}t+\dfrac{7}{3}e^{-t}+t-1=-\dfrac{7}{3}+\dfrac{4}{3}t+\dfrac{7}{3}e^{-t}$

$x(t)=-\dfrac{7}{3}t+\dfrac{2}{3}t^2-\dfrac{7}{3}e^{-t}$

8 Systems of Ordinary Differential Equations

Solve by Differential Operator and Laplace Transform.

Problem 801

Show that the solution to $\dfrac{dx}{dt} = 3e^{-t} \quad \dfrac{dy}{dt} = x + y$

is $x = -3e^{-t} + c_1 \quad y = \tfrac{3}{2}e^{-t} - c_1 + c_2 e^{t}$

Problem 802

Show that the solution to $\dfrac{dx}{dt} = e^{t} \quad \dfrac{dy}{dt} = \dfrac{x-y}{t}$

is $x = e^{t} + c_1 \quad y = t^{-1}e^{t} + c_1 + c_2 t^{-1}$

Problem 803

Show that the solution to $\dfrac{dx}{dt} = x + \sin t \quad \dfrac{dy}{dt} = t - y$

is $x = c_1 e^{t} - \tfrac{1}{2}(\sin t + \cos t) \quad y = t - 1 + c_2 e^{-t}$

Problem 804

Show that the solution to $\dfrac{d^2 x}{dt^2} + 4x = 3\sin t \quad \dfrac{dx}{dt} - \dfrac{d^2 y}{dt^2} + y = 2\cos t$

is $x = \sin t + c_1 \sin 2t + c_2 \cos 2t$

$y = \tfrac{1}{2}\cos t - \tfrac{2}{5}c_1 \cos 2t + \tfrac{2}{5}c_2 \sin 2t + c_3 e^{t} + c_4 e^{-t}$

Problem 805

Show that the solution to $\dfrac{d^2 x}{dt^2} - \dfrac{dy}{dt} = 1 - t \quad \dfrac{dx}{dt} + 2\dfrac{dy}{dt} = 4e^{t} + x$

is $x = c_1 e^{-t} + c_2 e^{t/2} + 2e^{t} + 2t \quad y = -c_1 e^{-t} + \tfrac{1}{2}c_2 e^{t/2} + c_3 + 2e^{t} - t + \tfrac{1}{2}t^2$

Problem 806

Show that the solution to $\dfrac{d^2 x}{dt^2} - x + \dfrac{d^2 y}{dt^2} + y = 0 \quad \dfrac{dx}{dt} + 2x + \dfrac{dy}{dt} + 2y = 0$

is $x = 5c_1 e^{-2t} \quad y = -3c_1 e^{-2t}$

Problem 807

Show that the solution to

$(D-1)x \qquad\qquad = 0$

$-x + (D-3)y \qquad = 0$

$-x + \qquad y + (D-2)z = 0$

is $x = c_1 e^{t} \quad y = c_2 e^{3t} - \tfrac{1}{2}c_1 e^{t} \quad z = c_3 e^{2t} - \tfrac{3}{2}c_1 e^{t} - c_2 e^{3t}$

Mathematics beyond the Calculus

9 Partial Differential Equations

Partial differential equations are of widespread interest because they are connected to many physical problems.

A partial differential equation subject to certain conditions in the form of *initial* or *boundary* conditions is known as an initial value or a boundary value problem. The initial conditions are the values of the unknown function f and an appropriate number of its derivatives, for example, at various values of the usual variables t and x.

Definitions: (1) $\dfrac{\partial f}{\partial t} = f_t \quad \dfrac{\partial^2 f}{\partial t^2} = f_{tt} \quad \dfrac{\partial f}{\partial x} = f_x \quad \dfrac{\partial^2 f}{\partial x^2} = f_{xx} \quad \dfrac{\partial^2 f}{\partial x \partial t} = f_{xt}$

The transform of the derivative is the derivative of the transform

(2) $\mathscr{L}\left[\dfrac{\partial f(x,t)}{\partial x}\right] = \int_0^\infty e^{-pt}\dfrac{\partial f(x,t)}{\partial x}dt = \dfrac{\partial}{\partial x}\int_0^\infty e^{-pt}f(x,t)dt = \dfrac{\partial}{\partial x}F(x,p)$

where $\mathscr{L}[f(x,t)] = \int_0^\infty e^{-pt}f(x,t)dt = F(x,p)$

Assume limits pass through the transform.

(3) $\lim\limits_{x \to x_0}\int_0^\infty e^{-pt}f(x,t)dt = \int_0^\infty e^{-pt}f(x_0,t)dt \to \int_0^\infty e^{-pt}f(x_0,t)dt = F(x_0,p)$

9.1 Solution by Separation of Variables

The basic idea of solution by separation of variables is to transform a partial differential equation of n variables into n ordinary differential equations. Then the solution to the partial differential equation is the *product* of the solutions of the ordinary differential equations.

This method solves a surprising number of partial differential equations.

Example

(4) $\dfrac{\partial^2 y(x,t)}{\partial x^2} - 2\dfrac{\partial y(x,t)}{\partial x} + \dfrac{\partial y(x,t)}{\partial t} = 0 \quad$ and assume $y = X(x) \times T(t)$

(5) let $\dfrac{dX}{dx} = X'(x) = X' \quad$ and $\quad \dfrac{dT}{dt} = T'(t) = T'$

9 Partial Differential Equations

Substituting (5) into (4) the variables are separated by using y=XT.

(6) $\dfrac{\partial^2 y(x,t)}{\partial x^2} - 2\dfrac{\partial y(x,t)}{\partial x} + \dfrac{\partial y(x,t)}{\partial t} = 0 \;\;\rightarrow\;\; X''T - 2X'T + XT' = 0$

$\dfrac{X''T - 2X'T}{XT} = -\dfrac{XT'}{XT} \;\;\rightarrow\;\; \dfrac{X''-2X'}{X} = -\dfrac{T'}{T}$

If x is fixed and t varies, then the left member is constant. And, if t is fixed and x varies, then the right member is constant. Consequently the common value of the two members is some number k, which results in two ordinary differential equations solved by methods in Chapter 7.

(7a) $\dfrac{X''-2X'}{X} = -\dfrac{T'}{T} \;\;\rightarrow\;\; \dfrac{X''-2X'}{X} = -\dfrac{T'}{T} = k$

(7b) $X''-2X'-kX = 0 \quad \text{and} \quad T'+kT = 0$

The auxiliary (or characteristic) equations are

(8) $m^2 - 2m - k = 0 \quad \text{and} \quad n + k = 0$

The roots are

(9) $m = 1 \pm \sqrt{1+k} \quad \text{and} \quad n = -k$

There are three cases: k > 0, k = 0, k < 0

(10)
k	X roots	X roots $k=3$	T roots $n=-k$
>0	$m = 1 \pm \sqrt{1+k}$	$m = 3, -1$	$n = -3$
=0	$m = 1 \pm 1$	$m = 1, 0$	$n = 0$
<0	$m = 1 \pm \sqrt{1-|k|}$	$m = 1+i\sqrt{2}, 1-i\sqrt{2}$	$n = 3$

Solutions (Section 7.2)

(11)
k	solution
>0	$y = c_1 e^{(3x-3t)} + c_2 e^{(-x-3t)}$
=0	$y = c_1 e^{(x)} + c_2$
<0	$y = \Re eal\; c_1 e^{(x+ix\sqrt{2}+3t)} + \Re eal\; c_2 e^{(x-ix\sqrt{2}+3t)}$

Constant coefficients c_1 and c_2 are computed from initial conditions, which have yet to be specified.

Initial and Boundary Conditions

A partial differential equation subject to certain conditions in the form of initial or boundary conditions is known as an initial value or a boundary value problem. The initial conditions are the values of the unknown function f and an appropriate number of its derivatives at the initial point.

There are three categories of boundary conditions.
1) *Dirichlet conditions* are the values of the unknown function u defined at each point of the boundary.

2) *Neumann conditions* are the values of the normal derivatives of the unknown function u defined at each point of the boundary.

3) *Mixed conditions* are the values of a linear combination of the unknown function f and its derivatives defined on each point of the boundary.

Examples

definitions $\quad \dfrac{\partial f}{\partial t} = f_t \quad \dfrac{\partial^2 f}{\partial t^2} = f_{tt} \quad \dfrac{\partial f}{\partial x} = f_x \quad \dfrac{\partial^2 f}{\partial x^2} = f_{xx}$

(1) $f_t = kf_{xx} \quad 0 < x < L, t > 0$
$\quad f(x,0) = g(x) \quad f_t(x,0) = h(x) \quad\quad 0 < x < L$
$\quad f(0,t) = t_1 \quad\quad f(L,t) = t_2 \quad\quad\quad t > 0$

(2) $f_t = kf_{xx} \quad 0 < x < L, t > 0$
$\quad f(x,0) = g(x) \quad f_t(x,0) = h(x) \quad\quad 0 < x < L$
$\quad f_x(0,t) = t_1 \quad\quad f_x(L,t) = t_2 \quad\quad\quad t > 0$

(3) $f_t = kf_{xx} \quad 0 < x < L, t > 0$
$\quad f(x,0) = g(x) \quad f_t(x,0) = h(x) \quad\quad 0 < x < L$
$\quad f(0,t) + af_x(0,t) = 0 \quad\quad t > 0$
$\quad f(L,t) + bf_x(L,t) = 0 \quad\quad t > 0$

9 Partial Differential Equations

9.2 Solution by Laplace Transform

Let $F(x,p)$ be the transform of $f(x, t)$ with respect to t.

(12) $F(x, p) = \mathcal{L}[f(x,t)] = \int_0^\infty e^{-pt} f(x,t) dt$

The method's steps
1. Transform t to p to get an ordinary differential equation (ODE) in p.
2a. Solve the ODE using solution by Laplace transform of x to λ (7.1)
2b. or Solve the ODE using solution by differential operator (7.2).
3a. If 2a perform 3a1 inverse transform λ to x, then 3a2 p to t.
3b. If 2b perform inverse transform p to t.

Example Step 1

(13a) $\dfrac{\partial f}{\partial x} = \dfrac{\partial f}{\partial t}$ and $f(x,0^+) = x$ $f(0,t) = t$ initial conditions

(13b) $\mathcal{L}_t\left[\dfrac{\partial f(x,t)}{\partial x}\right] = \mathcal{L}_t\left[\dfrac{\partial f(x,t)}{\partial t}\right]$ → $\dfrac{\partial}{\partial x} F(x, p) = pF(x, p) - f(x,0^+)$

(13c) ODE in p $\dfrac{d}{dx} F(x, p) - pF(x, p) = -x$

Steps 2a & 3a1 Solve the ordinary differential equation. Transform x to λ

(14) step 2a - transform x to λ → $\mathcal{L}_\lambda\left[\dfrac{d}{dx} F(x, p)\right] - p\mathcal{L}_\lambda[F(x, p)] = -\mathcal{L}_\lambda[x]$

$\lambda \mathcal{L}_\lambda[F(x, p)] - F(0, p) - p\mathcal{L}_\lambda[F(x, p)] = -\dfrac{1}{\lambda^2}$

$\mathcal{L}_\lambda[F(x, p)] = \dfrac{1}{\lambda - p} F(0, p) - \dfrac{1}{\lambda - p} \dfrac{1}{\lambda^2}$

$= \dfrac{1}{\lambda - p} F(0, p) + \dfrac{1}{\lambda p^2} + \dfrac{1}{\lambda^2 p} - \dfrac{1}{p^2} \dfrac{1}{\lambda - p}$

step 3a1 - inverse transform λ to x - $F(x, p) = F(0, p)e^{px} + \dfrac{1}{p^2} + \dfrac{x}{p} - \dfrac{1}{p^2} e^{px}$

Step 3a2 Inverse transform p to t.

(15a) $F(0, p) = \mathcal{L}[f(0,t)] = \int_0^\infty e^{-pt} f(0,t) dt = \int_0^\infty e^{-pt} t\, dt = \dfrac{1}{p^2}$

(15b) $F(x, p) = \dfrac{1}{p^2} e^{px} + \dfrac{1}{p^2} + \dfrac{x}{p} - \dfrac{1}{p^2} e^{px} = \dfrac{1}{p^2} + \dfrac{x}{p}$

(15c) $f(x,t) = (t + x)u(t)$

Mathematics beyond the Calculus

Example – vibrating string

(16a) $\dfrac{\partial^2 y}{\partial t^2} = a^2 \dfrac{\partial^2 y}{\partial x^2}$ with initial conditions string at rest

$y(x,0^+) = y_t(x,0^+) = 0 \quad y(0,t) = f(t) \ [f(0) = 0] \quad \lim\limits_{x \to \infty} y(x,t) = 0$

(16b) $\mathcal{L}_{tt}\left[\dfrac{\partial^2 y(x,t)}{\partial t^2}\right] = a^2 \mathcal{L}_{xx}\left[\dfrac{\partial^2 y(x,t)}{\partial x^2}\right]$

$p^2 Y(x,p) - py(x,0^+) - y_t(x,0^+) = a^2 \dfrac{\partial^2 Y(x,p)}{\partial x^2}$

$p^2 Y(x,p) - p0 - 0 = a^2 \dfrac{\partial^2 Y(x,p)}{\partial x^2} \rightarrow \text{ODE} \quad \dfrac{\partial^2 Y(x,p)}{\partial x^2} - \dfrac{p^2}{a^2} Y(x,p) = 0$

by method of 7.2 $\left(D^2 - \dfrac{p^2}{a^2}\right) Y(x,p) = 0 \rightarrow Y(x,p) = c_1 e^{\frac{p}{a}x} + c_2 e^{-\frac{p}{a}x}$

Invoke initial conditions to evaluate c_1 and c_2.

(16c) $\lim\limits_{x \to \infty} y(x,t) = 0 \rightarrow \lim\limits_{x \to \infty} Y(x,0) = 0 \rightarrow c_1 = 0 \rightarrow Y(x,p) = c_2 e^{-\frac{p}{a}x}$

$y(0,t) = f(t) \ [f(0) = 0]$

$Y(0,p) = \mathcal{L}_t[y(0,t)] = \mathcal{L}_t[f(t)]$

$Y(0,p) = c_2 e^{-\frac{p}{a}0} = c_2 = \mathcal{L}_t[f(t)] \rightarrow Y(x,p) = \mathcal{L}_t[f(t)] e^{-\frac{p}{a}x}$

$\rightarrow y(x,t) = f\!\left(t - \dfrac{x}{a}\right) u\!\left(t - \dfrac{x}{a}\right)$

Example – heat equation

(17a) $\dfrac{\partial^2 u}{\partial x^2} = \dfrac{\partial u}{\partial t}$ and $u(x,0^+) = 1 \ x > 0 \quad u(0,t) = 0 \ t > 0 \quad \lim\limits_{x \to \infty} u(x,t) = 1$

(17b) $\mathcal{L}_t\left[\dfrac{\partial^2 u(x,t)}{\partial x^2}\right] = \mathcal{L}_t\left[\dfrac{\partial u(x,t)}{\partial t}\right]$

$\dfrac{d^2 U(x,p)}{dx^2} = pU(x,p) - u(x,0^+) \rightarrow \text{ODE} \quad \dfrac{d^2 U(x,p)}{dx^2} = pU(x,p) - 1$

by 7.2 $(D^2 - p)U(x,p) = 1 \rightarrow U(x,p) = c_1 e^{\sqrt{p}\,x} + c_2 e^{-\sqrt{p}\,x} + \dfrac{1}{p}$

(17c) $\lim\limits_{x \to \infty} u(x,t) = 1 \rightarrow \lim\limits_{x \to \infty} U(x,p) = \dfrac{1}{p} \rightarrow c_1 = 0 \rightarrow U(x,p) = c_2 e^{-\sqrt{p}\,x} + \dfrac{1}{p}$

$u(0,t) = 0 \rightarrow U(0,p) = 0 \rightarrow 0 = c_2 + \dfrac{1}{p} \rightarrow c_2 = -\dfrac{1}{p}$

$\rightarrow U(x,p) = \dfrac{1}{p} - \dfrac{1}{p} e^{-\sqrt{p}\,2(x/2)} \rightarrow y(x,t) = 1 - \text{erfc}\!\left(\dfrac{x}{2\sqrt{t}}\right) = \text{erf}\!\left(\dfrac{x}{2\sqrt{t}}\right)$

9 Partial Differential Equations

Problem 901 Wave Equation

Solve $\dfrac{\partial^2 f(x,t)}{\partial x^2} = \dfrac{\partial^2 f(x,t)}{\partial t^2}$ i.e. $f_{xx}(x,t) = f_{tt}(x,t)$ subject to

$f(0,t) = f(t)$ $t > 0$ $f(0) = 0$ $f(x,0) = 0$ $f_t(x,0) = 0$ $x > 0$ $\lim\limits_{x \to \infty} f(x,t) = 0$

where $F(x,p) = \mathcal{L}[f(x,t)] = \int_0^\infty e^{-pt} f(x,t)\,dt$

Hints
Step 1 Show that the ODE subsiduary equation is

$p^2 F(x,p) - pf(x,0) - f_t(x,0) = \dfrac{\partial^2 F(x,p)}{\partial x^2} \;\to\; \dfrac{\partial^2 F(x,p)}{\partial x^2} - p^2 F(x,p) = 0$

$\to$ ODE $\dfrac{d^2 F(x,p)}{dx^2} - p^2 F(x,p) = 0$

Step 2b by method of 7.2 show that $F(x,p) = c_1 e^{px} + c_2 e^{-px}$

Step 3b Show that $f(x,t) = \begin{cases} 0 & \text{if } x > t \\ f(t-x) & \text{if } x \le t \end{cases}$

Problem 902

Solve $\dfrac{\partial^2 f(x,y)}{\partial x^2} + \dfrac{\partial^2 f(x,y)}{\partial x \partial y} + \dfrac{\partial^2 f(x,y)}{\partial y^2}$ subject to $f(x,0) = 0$ $f_t(x,0) = 0$ $x > 0$

where $F(x,p) = \mathcal{L}[f(x,y)] = \int_0^\infty e^{-py} f(x,y)\,dy$

Hints
Step 1 Show that the ODE subsiduary equation is

$\dfrac{\partial^2 F(x,p)}{\partial x^2} + p\dfrac{\partial F(x,p)}{\partial x} - f(x,0) + p^2 F(x,p) - pf(x,0) - f_t(x,0) = 0$

$\dfrac{\partial^2 F(x,p)}{\partial x^2} + p\dfrac{\partial F(x,p)}{\partial x} + p^2 F(x,p) = 0$

ODE $\dfrac{d^2 F(x,p)}{dx^2} + p\dfrac{dF(x,p)}{dx} + p^2 F(x,p) = 0 \;\to\; (D^2 + pD + p^2)F(x,p) = 0$

Step 2b by method of 7.2 show that $F(x,p) = c_1 e^{\frac{1}{2}(-1+i\sqrt{3})px} + c_2 e^{\frac{1}{2}(-1-i\sqrt{3})px}$

Mathematics beyond the Calculus

Problem 903

Solve $\dfrac{\partial^2 u}{\partial t^2} - \dfrac{\partial^2 u}{\partial x^2} - u = 0$ for $0 < x < 1,\ t > 0$

subject to $\dfrac{\partial u(x,0)}{\partial t} = 0 \quad u(0,t) = u(1,t) = 0$

Show that the solution is $u(x,t) = \cos at\ \sin n\pi x \quad$ where $a = \sqrt{n^2 \pi^2 - 1}$

Problem 904

Solve $\dfrac{\partial^2 u}{\partial t^2} + 2\dfrac{\partial u}{\partial t} - 4\dfrac{\partial^2 u}{\partial x^2} + u = 0$ for $0 < x < 1,\ t > 0$

subject to $u(x,0) = 0 \quad \dfrac{\partial u(0,t)}{\partial x} = 0 \quad u(1,t) = 0$

Show that the solution is $u(x,t) = e^{-t} \sin(2n-1)\pi t\ \cos\left(n - \tfrac{1}{2}\right)\pi x \quad n = 1,2,\ldots$

Problem 905

Solve $\dfrac{\partial u}{\partial t} - t^2 \dfrac{\partial^2 u}{\partial x^2} - u = 0$ for $0 < x < 1,\ t > 0$

subject to $u(0,t) = u(1,t) = 0$

Show that the solution is $u(x,t) = e^{t - at^3} \sin n\pi x \quad$ where $a = \tfrac{1}{3} n^2 \pi^2$

Problem 906

Solve $\dfrac{\partial u}{\partial t} - \dfrac{\partial^2 u}{\partial x^2} - 2\dfrac{\partial u}{\partial x} = 0$ for $1 < x < 2,\ t > 0$

subject to $u(1,t) = 0 \quad u(2,t) = 0$

Show that the solution is $u(x,t) = e^{n^2 \pi^2 t - x} \sin n\pi(x-1) \quad n = 1,2,\ldots$

Problem 907

Solve $\dfrac{\partial^2 u}{\partial x^2} + \dfrac{\partial^2 u}{\partial y^2} + \dfrac{\partial u}{\partial x} = 0$ for $0 < x < \pi$ and $0 < y < \pi$

subject to $u(x,0) = 0 \quad u(x,\pi) = 0 \quad u(0,y) = 0 \quad u(\pi,y) = \sin y$

Show that the solution is $u(x,y) = e^{(\pi - x)/2}\ \dfrac{\sinh \sqrt{5}\,x/2}{\sinh \sqrt{5}\,\pi/2} \sin y$

10 Fourier Series

The trigonometrical series
(1) $f(x) = (a_0/2) + (a_1 \cos x + b_1 \sin x) + (a_2 \cos 2x + b_2 \sin 2x) + ...$
is said to be a *Fourier* series when the coefficients a_0, a_1, b_1, a_2, b_2, etc are given by

(2a) $a_0 = \dfrac{1}{\pi} \int_c^{c+2\pi} f(x)dx$

and when $n \geq 1$

(2b) $a_n = \dfrac{1}{\pi} \int_c^{c+2\pi} f(x)\cos nx \, dx$ (2c) $b_n = \dfrac{1}{\pi} \int_c^{c+2\pi} f(x)\sin nx \, dx$

If both sides of equation (1) are multiplied by *cos nx* and integrated from $-\pi$ to π, then equation (3b) produces equation (2b).

(3a) $\int_c^{c+2\pi} f(x)\cos nx \, dx = \int_c^{c+2\pi} a_0 \cos nx \, dx + \int_c^{c+2\pi} a_1 \cos x \cos nx \, dx$
$+ \int_c^{c+2\pi} b_1 \sin x \cos nx \, dx + ... + \int_c^{c+2\pi} a_n \cos nx \cos nx \, dx + \int_c^{c+2\pi} b_n \sin nx \cos nx \, dx + ...$

(3b) $\int_c^{c+2\pi} f(x)\cos nx \, dx = (a_0 0) + (a_1 0) + (b_1 0) + ... + (a_n \pi) + (b_n 0) + ...$

If both sides of equation (1) are multiplied by *sin nx* and integrated from $-\pi$ to π, then equation (4b) produces equation (2c).

(4a) $\int_c^{c+2\pi} f(x)\sin nx \, dx = \int_c^{c+2\pi} a_0 \sin nx \, dx + \int_c^{c+2\pi} a_1 \cos x \sin nx \, dx$
$+ \int_c^{c+2\pi} b_1 \sin x \sin nx \, dx + ... + \int_c^{c+2\pi} a_n \cos nx \sin nx \, dx + \int_c^{c+2\pi} b_n \sin nx \sin nx \, dx + ...$

(4b) $\int_c^{c+2\pi} f(x)\sin nx \, dx = (a_0 0) + (a_1 0) + (b_1 0) + ... + (a_n 0) + (b_n \pi) + ...$

Fourier's Theorem Any single valued function f(x), continuous except possibility for a finite number of finite discontinuities in an interval of length 2π, and having only a finite number of maxima and minima in this interval, possesses a convergent Fourier series representing it.

Note: It is not true that all continuous functions can be represented by a Fourier series.

Mathematics beyond the Calculus

Even and odd functions
A function f(x) is said to be *even* when $f(-x) \equiv f(x)$. The Fourier series of even functions are a sum of cosines. A function f(x) is said to be *odd* when $f(-x) = -f(x)$. The Fourier series of odd functions are a sum of sines.

One need not be concerned about this, because even functions will produce coefficients $b_n = 0$ and odd functions will produce coefficients $a_n = 0$.

Change of Interval In many applications an interval different from $-\pi$ to π is required. Consider any 2L interval from $-L$ to L.

(5a) if $\dfrac{y}{x} = \dfrac{L}{\pi}$ then $x = \dfrac{\pi y}{L}$ and $y = \dfrac{Lx}{\pi}$

(5b) $a_0 = \dfrac{1}{\pi}\int_c^{c+2\pi} f(x)dx = \dfrac{1}{\pi}\int_{-L}^{L} f\left(\dfrac{\pi y}{L}\right)\dfrac{\pi}{L}dy = \dfrac{1}{L}\int_{-L}^{L} g(y)dy$

and when $n \geq 1$

(5c) $a_n = \dfrac{1}{\pi}\int_c^{c+2\pi} f(x)\cos nx\, dx$

$= \dfrac{1}{\pi}\int_{-L}^{L} f\left(\dfrac{\pi y}{L}\right)\cos\dfrac{n\pi y}{L}\dfrac{\pi}{L}dy = \dfrac{1}{L}\int_{-L}^{L} g(y)\cos\dfrac{n\pi y}{L}dy$

(5d) $b_n = \dfrac{1}{L}\int_{-L}^{L} g(y)\sin\dfrac{n\pi y}{L}dy$

Example 1
(6a) $g(y) = 0 \quad -2 < y < 0 \quad$ and $\quad g(y) = b \quad 0 < y < 2$

(6b) $a_0 = \dfrac{1}{L}\int_{-L}^{L} g(y)dy = \dfrac{1}{2}\int_{-2}^{0} 0\, dy + \dfrac{1}{2}\int_{0}^{2} b\, dy = b$

and when $n \geq 1$

(6c) $a_n = \dfrac{1}{L}\int_{-L}^{L} g(y)\cos\dfrac{n\pi y}{L}dy = \dfrac{1}{2}\int_{-2}^{0} 0\cos\dfrac{n\pi y}{2}dy + \dfrac{1}{2}\int_{0}^{2} b\cos\dfrac{n\pi y}{2}dy$

$= 0 + \left[\dfrac{b}{n\pi}\sin\dfrac{n\pi y}{2}\right]_0^2 = 0 + 0 - 0 = 0$

(6d) $b_n = 0 + \left[-\dfrac{b}{n\pi}\cos\dfrac{n\pi y}{2}\right]_0^2 = \dfrac{b}{n\pi}(1 - \cos n\pi)$

(6e) $g(y) = \dfrac{b}{2} + \dfrac{2b}{\pi}\left(\sin\dfrac{\pi y}{2} + \dfrac{1}{3}\sin\dfrac{3\pi y}{2} + \dfrac{1}{5}\sin\dfrac{5\pi y}{2} + ...\right)$

9 Fourier Series

Example 2

(7a) $L = \pi$ $f(x) = -x$ $-\pi < x \le 0$ and $f(x) = x$ $0 < x < \pi$

(7b) $a_0 = \dfrac{1}{L}\int_{-L}^{L} f(x)dx = \dfrac{1}{\pi}\int_{-\pi}^{0} -x\,dx + \dfrac{1}{\pi}\int_{0}^{\pi} x\,dx = \dfrac{\pi}{2} + \dfrac{\pi}{2} = \pi$

and when $n \ge 1$

(7c) $a_n = \dfrac{1}{L}\int_{-L}^{L} f(x)\cos\dfrac{n\pi x}{L}dx = -\dfrac{1}{\pi}\int_{-\pi}^{0} x\cos nx\,dx + \dfrac{1}{\pi}\int_{0}^{\pi} x\cos nx\,dx$

integrate by parts $u = x$ $du = dx$ $dv = \cos nx\,dx$ $v = \dfrac{1}{n}\sin nx$

$\int u\,dv = uv - \int v\,du \;\to\; \int x\cos nx\,dx = x\dfrac{1}{n}\sin nx - \int \dfrac{1}{n}\sin nx\,dx$

$= -\dfrac{1}{\pi}\left[\dfrac{x}{n}\sin nx + \dfrac{1}{n^2}\cos nx\right]_{-\pi}^{0} + \dfrac{1}{\pi}\left[\dfrac{x}{n}\sin nx + \dfrac{1}{n^2}\cos nx\right]_{0}^{\pi}$

$= -\dfrac{1}{\pi}\left[0 + \dfrac{1}{n^2} - 0 + \dfrac{1}{n^2}\cos n\pi\right] + \dfrac{1}{\pi}\left[0 + \dfrac{1}{n^2}\cos n\pi - 0 - \dfrac{1}{n^2}\cos n0\right]$

$= -\dfrac{1}{\pi n^2} + \dfrac{1}{\pi n^2}\cos n\pi + \dfrac{1}{\pi n^2}\cos n\pi - \dfrac{1}{\pi n^2} = \dfrac{2}{\pi n^2}(\cos n\pi - 1)$

(7d) $b_n = \dfrac{1}{L}\int_{-L}^{L} f(x)\sin\dfrac{n\pi x}{L}dx = -\dfrac{1}{\pi}\int_{-\pi}^{0} x\sin nx\,dx + \dfrac{1}{\pi}\int_{0}^{\pi} x\sin nx\,dx$

$= -\dfrac{1}{\pi}\left[-\dfrac{x}{n}\cos nx + \dfrac{1}{n^2}\sin nx\right]_{-\pi}^{0} + \dfrac{1}{\pi}\left[-\dfrac{x}{n}\cos nx + \dfrac{1}{n^2}\sin nx\right]_{0}^{\pi}$

$= -\dfrac{1}{\pi}\left[0 + 0 - \dfrac{\pi}{n}\cos n\pi - 0\right] + \dfrac{1}{\pi}\left[-\dfrac{\pi}{n}\cos n\pi + 0 + 0 + 0\right]$

$= \dfrac{1}{n}\cos n\pi - \dfrac{1}{n}\cos n\pi = 0$

(7e) $g(y) = \dfrac{b}{2} + \dfrac{2b}{\pi}\left(\sin\dfrac{\pi y}{2} + \dfrac{1}{3}\sin\dfrac{3\pi y}{2} + \dfrac{1}{5}\sin\dfrac{5\pi y}{2} + ...\right)$

Example 3

(8a) $L = \pi$ $f(x) = x$ $-\pi < x < \pi$

(8b) $a_0 = \dfrac{1}{L}\int_{-L}^{L} f(x)dx = \dfrac{1}{\pi}\int_{-\pi}^{\pi} x\,dx = 0$

and when $n \ge 1$

(8c) $a_n = \dfrac{1}{L}\int_{-L}^{L} f(x)\cos\dfrac{n\pi x}{L}dx = \dfrac{1}{\pi}\int_{-\pi}^{\pi} x\cos nx\,dx = 0$

(8d) $b_n = \dfrac{1}{L}\int_{-L}^{L} f(x)\sin\dfrac{n\pi x}{L}dx = \dfrac{1}{\pi}\int_{-\pi}^{\pi} x\sin nx\,dx = -\dfrac{2}{n}\cos n\pi$

(8e) $f(x) = 2\left(\sin x - \dfrac{1}{2}\sin 2x + \dfrac{1}{3}\sin 3x - ....\right)$

Mathematics beyond the Calculus

Problem 1001

Show that $\int_{-\pi}^{\pi} \sin nx \, dx = 0$ $\int_{-\pi}^{\pi} \cos nx \, dx = 0$

Problem 902

Show that $\int_{-\pi}^{\pi} \cos mx \cos nx \, dx = \begin{cases} 0 & m \neq n \\ \pi & m = n \end{cases}$

hint $2\cos a \cos b = \cos(a+b) + \cos(a-b)$

repeat for $\int_{-\pi}^{\pi} \cos mx \sin nx \, dx$, $\int_{-\pi}^{\pi} \sin mx \sin nx \, dx$

Problem 1003
Show that the Fourier series of $\cos nx$ and $\sin nx$ for
$f(x) = 0 \quad -\pi < x \leq 0$ and $f(x) = \frac{\pi}{4} x \quad 0 < x < \pi$

has coefficients $a_0 = \frac{\pi^2}{16}$ $a_n = \frac{1}{4n^2}(\cos n\pi - 1)$ $b_n = -\frac{\pi}{4n} \cos n\pi$

and show that when $x = \pm\pi$ $\dfrac{\pi^2}{8} = 1 + \dfrac{1}{3^2} + \dfrac{1}{5^2} + \ldots$

Problem 1004
Show that the Fourier series of $\cos nx$ and $\sin nx$ for $f(x) = x + x^2$ $-\pi < x < \pi$

has coefficients $a_0 = \frac{\pi^2}{3}$ $a_n = \frac{4}{n^2} \cos n\pi$ $b_n = (-1)^{n-1} \frac{2}{n}$

hint integrate by parts

and show that when $x = \pm\pi$ $\dfrac{\pi^2}{6} = 1 + \dfrac{1}{2^2} + \dfrac{1}{3^2} + \ldots$

Problem 1005
Show that the Fourier series of $\cos nx$ and $\sin nx$ for

$f(x) = \frac{\pi}{2\sinh \pi} e^x \quad -\pi < x < \pi$ hint integrate by parts

has coefficients $a_0 = \frac{1}{2}$ $a_n = \frac{(-1)^n}{n^2+1}$ $b_n = \frac{n}{n^2+1}(-1)^{n-1}$

$f(x) = \frac{1}{2} + \left(\frac{-1}{1^2+1} \cos x + \frac{1}{1^2+1} \sin x + \frac{1}{2^2+1} \cos 2x + \frac{-2}{2^2+1} \sin 2x + \ldots \right)$

Problem 1006
Show that the Fourier series of $\cos nx$ for

$f(x) = 0 \quad 0 \leq x < \frac{\pi}{2}$ $f(\frac{\pi}{2}) = \frac{\pi}{4}$ and $f(x) = \frac{\pi}{2} \quad \frac{\pi}{2} < x \leq \pi$

has coefficients $a_0 = \frac{\pi}{4}$ $a_n = -\frac{1}{n}\sin\frac{n\pi}{2}$ $b_n = 0$

$f(x) = \frac{\pi}{4} - \left(\cos x - \frac{1}{3}\cos 3x + \frac{1}{5}\cos 5x + \ldots \right)$

11 The Z Transform

The Z Transform is the basis for a straightforward process that solves *difference equations*. Difference equations are expressions representing relations amongst sequences of numbers x(n) *where n is an integer.*

Difference equation solutions start by transforming the equations into subsidiary equations, which are algebraic equations. The subsidiary equations are solved for the variables of interest by algebraic manipulations that may include a partial fraction expansion. Here initial conditions are invoked. The solved subsidiary equations are inverse transformed back to the original domain *as a solution* of the original problem.

11.1 Z Transform Defined

The Laplace transform of f(t) is defined in Chapter 4 as

(1) $\quad F(p) = \mathcal{L}[f(t)] = \int_0^\infty f(t)e^{-pt}\,dt$

If we use the Laplace transform to solve a problem in the *complex frequency p* domain, then we need an inverse transform to return to the t domain.

(2) $\quad f(t) = \mathcal{L}^{-1}[F(p)] = \dfrac{1}{2\pi i}\int_{\sigma-j\infty}^{\sigma+j\infty} F(p)e^{tp}\,dp$

On the other hand the Z transform is defined as a sum instead of an integral in order to deal with sequential functions of n, which is an integer variable.

(3a) $\quad F(z) = Z[f(n)] = \sum_{n=0}^{\infty} f(n)z^{-n} = f(0) + f(1)z^{-1} + f(2)z^{-2} + f(3)z^{-3} + \ldots$

(3b) *inverse transform* $\quad f(n) = \dfrac{1}{2\pi i}\oint F(z)z^{n-1}\,dz$

Define the sequence f(n) as the sum of a sequence of numbers f(k) *where n and k are integers.*

(4) $\quad f(n) = \sum_{k=0}^{\infty} f(k)\delta(n-k)$

Mathematics beyond the Calculus

The function δ(n) is analogous to the Dirac delta function (Figure 1101).

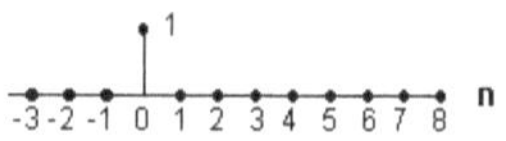

(5) $\delta(n) = \begin{cases} 1 & n = 0 \\ 0 & n \neq 0 \end{cases}$

Figure 1101 δ(n)

The Z transform Z[f(n)] may be derived from the Laplace transform.

(6a) *sample a continuous time signal* $f(t)$ *to produce* $f_s(t)$

$$f_s(t) = f(t)\sum_{n=0}^{\infty} \delta(t - nT) = \sum_{n=0}^{\infty} f(nT)\delta(t - nT)$$

(6b) *the Laplace transform of* $f_s(t)$ *is*

$$\mathcal{L}[f_s(t)] = \int_0^{\infty} f_s(t)e^{-pt} dt = \int_0^{\infty} \sum_{n=0}^{\infty} f(nT)\delta(t - nT)e^{-pt} dt$$

$$= \sum_{n=0}^{\infty} \int_0^{\infty} f(nT)\delta(t - nT)e^{-pt} dt = \sum_{n=0}^{\infty} f(nT) \int_0^{\infty} \delta(t - nT)e^{-pt} dt$$

$$= \sum_{n=0}^{\infty} f(nT)e^{-pnT} = \sum_{n=0}^{\infty} f(nT)z^{-n} \quad \text{where } z = e^{pT}$$

let $f(n) = f(nT)$ then $\mathcal{L}[f_s(t)] = \sum_{n=0}^{\infty} f(n)z^{-n} = F(z) = Z[f(n)]$

Emphasis - The Z transform Z[f(n)] is a polynomial in z^{-1}.

(3) $F(z) = Z[f(n)] = \sum_{n=0}^{\infty} f(n)z^{-n} = f(0) + f(1)z^{-1} + f(2)z^{-2} + f(3)z^{-3} + \ldots$

11 The Z Transform

11.2 General Z Transforms

Transform of multiplication by a constant c
(7a) if $f(n) \Leftrightarrow F(z)$, then

$$Z[c \times f(n)] = \sum_{n=0}^{\infty} \{c \times f(n)\} z^{-n} = c \times \sum_{n=0}^{\infty} \{f(n)\} z^{-n} = c \times F(z)$$

(7b) $cf(n) \Leftrightarrow cF(z)$ (the $\Leftrightarrow$ symbol means transform from t to p or p to t)

Transform of a linear sum with constants c_1 and c_2
(8a) if $f_i(n) \Leftrightarrow F_i(z)$ $i = 1, 2$ then $Z[c_1 f_1(n) + c_2 f_2(n)]$

$$= Z[c_1 f_1(n)] + Z[c_2 f_2(n)] = \sum_{n=0}^{\infty} \{c_1 f_1(n)\} z^{-n} + \sum_{n=0}^{\infty} \{c_2 f_2(n)\} z^{-n} = c_1 F_1(z) + c_2 F_2(z)$$

(8b) $c_1 f_1(n) + c_2 f_2(n) \Leftrightarrow c_1 F_1(z) + c_2 F_2(z)$

Transform of a delay (#1) – a change in variable n to n – k
(9a) if $f(n) \Leftrightarrow F(z)$ then

$$Z[f(n-k)u(n-k)] = \sum_{n=0}^{\infty} \{f(n-k)u(n-k)\} z^{-n} = \sum_{n=0}^{\infty} \{f(n-k)\} z^{-n}$$

$$\text{if } m = n - k \text{ then } = \sum_{n=0}^{\infty} f(m) z^{-(m+k)} = z^{-k} \sum_{n=0}^{\infty} f(m) z^{-m} = z^{-k} F(z)$$

(9b) $f(n-k)u(n-k) \Leftrightarrow z^{-k} F(z)$

Transform of a delay (#2) – a change in variable n to n – k
(10a) if $f(n) \Leftrightarrow F(z)$ then

$$Z[f(n-k)u(n)] = \sum_{n=0}^{\infty} f(n-k) z^{-n} = z^{-k} \sum_{n=0}^{\infty} f(n-k) z^{-(n-k)} = z^{-k} \sum_{m=-k}^{\infty} f(m) z^{-m}$$

$$= z^{-k} \left(\sum_{m=0}^{\infty} f(m) z^{-m} + \sum_{m=-k}^{-1} f(m) z^{-m} \right) = z^{-k} \left(\sum_{m=0}^{\infty} f(m) z^{-m} + \sum_{m=k}^{1} f(-m) z^{m} \right)$$

where $m \to -m$ so that $Z[f(n-k)] = z^{-k} \left[F(z) + \sum_{m=1}^{k} f(-m) z^{m} \right]$

(10b) $f(n-k) \Leftrightarrow z^{-k} \left[F(z) + \sum_{m=1}^{k} f(-m) z^{m} \right]$ this form shows initial conditions

Examples
(10c) $f(n-1) \Leftrightarrow z^{-1} F(z) + f(-1)$
(10d) $f(n-3) \Leftrightarrow z^{-3} \left[F(z) + z^1 f(-1) + z^2 f(-2) + z^3 f(-3) \right]$

Mathematics beyond the Calculus

Derivative in the Z domain of a transform F(z)
(11a) if $f(n) \Leftrightarrow F(z)$, then

$$F(z) = Z[f(n)] = \sum_{n=0}^{\infty} f(n)z^{-n}$$

$$-z\frac{dF(z)}{dz} = -z\sum_{n=0}^{\infty} f(n)(-n)z^{-n-1} = n\sum_{n=0}^{\infty} f(n)z^{-n} = Z[nf(n)]$$

(11b) $\quad nf(n) \Leftrightarrow -z\dfrac{dF(z)}{dz}$

Convolution
(12a) if $f_i(n) \Leftrightarrow F_i(z)$ $i = 1,2$ and $f(n) = f_1(n) \times f_2(n)$ then

$$F(z) = Z[f(n)] = Z\left[\sum_{n=0}^{\infty} f(n)\right] = Z\left[\sum_{k=0}^{\infty} f_1(n) \times f_2(n-k)\right]$$

$$= \sum_{n=0}^{\infty}\left[\sum_{k=0}^{\infty} f_1(n) \times f_2(n-k)\right]z^{-n}$$

$$= \sum_{k=0}^{\infty} f_1(k)\left[\sum_{n=0}^{\infty} f_2(n-k)z^{-n}\right] = \sum_{k=0}^{\infty} f_1(k)z^{-k} F_2(z) = F_1(z)F_2(z)$$

(12b) $f_1(n) \times f_2(n) \Leftrightarrow F_1(z)F_2(z)$

Transform of an advance – a change in variable n to n + k
(13a) if $f(n) \Leftrightarrow F(z)$ then

$$Z[f(n+k)u(n)] = \sum_{n=0}^{\infty} f(n+k)z^{-n} = z^k \sum_{n=0}^{\infty} f(n+k)z^{-(n+k)} = z^{-k}\sum_{m=k}^{\infty} f(m)z^{-m}$$

$$= z^{-k}\left[\sum_{m=0}^{\infty} f(m)z^{-m} - \sum_{m=0}^{k-1} f(m)z^{-m}\right]$$

(13b) $f(n+k) \Leftrightarrow z^{-k}\left[F(z) - \sum_{m=0}^{k-1} f(m)z^{-m}\right]$ this form shows initial conditions

Initial value theorem
(14) if $f(n) = 0$ for $n < 0$ then $\lim_{z \to \infty} Z[f(n)] = f(0)$

11.3 Specific Z Transforms

Transform of $f(n)=\delta(n-a)$
(15a) $f(n) = \delta(n-a)$

$$F(z) = Z[f(n)] = \sum_{k=0}^{\infty} f(k)z^{-k} = \sum_{k=0}^{\infty} \delta(k-a)z^{-k} = z^{-a} \quad \text{where } \delta(0) = 1$$

(15b) $\delta(n-a) \Leftrightarrow z^{-a}$

Transform of $a^n f(n)$ scaling in the z domain
(16a) $g(n) = a^n f(n)$

$$F(z) = Z[g(n)] = \sum_{n=0}^{\infty} a^n f(n) z^{-n} = \sum_{k=0}^{\infty} f(n)\left(\frac{z}{a}\right)^{-n} = F\left(\frac{z}{a}\right)$$

(16b) $a^n f(n) \Leftrightarrow F\left(\dfrac{z}{a}\right)$

Transform of unit step $f(n)=u(n)$
(17a) $f(n) = u(n)$

$$F(z) = Z[f(n)] = \sum_{k=0}^{\infty} f(k)z^{-k} = \sum_{k=0}^{\infty} u(k)z^{-k} = \sum_{k=0}^{\infty} (z^{-1})^k = \frac{1}{1-z^{-1}} = \frac{z}{z-1}$$

(17b) $u(n) \Leftrightarrow \dfrac{z}{z-1} = \dfrac{1}{1-z^{-1}}$

Transform of $f(n)= b^n u(n)$
(18a) $f(n) = b^n u(n)$

$$F(z) = Z[f(n)] = \sum_{k=0}^{\infty} f(k)z^{-k} = \sum_{k=0}^{\infty} b^k z^{-k} = \sum_{k=0}^{\infty} \left(\frac{b}{z}\right)^k = \frac{1}{1-b/z} = \frac{z}{z-b}$$

(18b) $b^n u(n) \Leftrightarrow \dfrac{z}{z-b} = \dfrac{1}{1-bz^{-1}}$

Transform of $f(n)= e^{an} u(n)$
(19a) $f(n) = e^{an} u(n)$

$$F(z) = Z[f(n)] = \sum_{k=0}^{\infty} f(k)z^{-k} = \sum_{k=0}^{\infty} e^{ak} z^{-k} = \sum_{k=0}^{\infty} \left(\frac{e^a}{z}\right)^k = \frac{1}{1-e^a/z} = \frac{z}{z-e^a}$$

(19b) $e^{an} u(n) \Leftrightarrow \dfrac{z}{z-e^a} = \dfrac{1}{1-e^a z^{-1}}$

Mathematics beyond the Calculus

Transform of $f(n)= \cos(an)$

(20a) $f(n) = \cos an$

$F(z) = Z[f(n)] = Z[\cos an] = Z\left[\frac{1}{2}e^{ian} + \frac{1}{2}e^{-ian}\right] = Z\left[\frac{1}{2}e^{ian}\right] + Z\left[\frac{1}{2}e^{-ian}\right]$

$= \frac{1}{2}\frac{z}{z-e^{ia}} + \frac{1}{2}\frac{z}{z-e^{-ia}} = \frac{z}{2}\left(\frac{1}{z-e^{ia}} + \frac{1}{z-e^{-ia}}\right) = \frac{z}{2}\left(\frac{2z - e^{ia} - e^{-ia}}{(z-e^{ia})(z-e^{-ia})}\right)$

$= z\left(\frac{z-\cos a}{(z-e^{ia})(z-e^{-ia})}\right) = z\frac{z-\cos a}{z^2 - z(e^{ia}+e^{-ia})+1} = z\frac{z-\cos a}{z^2 - 2z\cos a + 1}$

(20b) $\cos an \Leftrightarrow z\dfrac{z-\cos a}{z^2 - 2z\cos a + 1} = \dfrac{1-(\cos a)z^{-1}}{1-2(\cos a)z^{-1}+z^{-2}}$

Transform of $f(n)= \sin(an)$

(21a) $f(n) = \sin an$

$F(z) = Z[f(n)] = Z[\sin an] = Z\left[\frac{1}{2}e^{ian} - \frac{1}{2}e^{-ian}\right] = Z\left[\frac{1}{2}e^{ian}\right] - Z\left[\frac{1}{2}e^{-ian}\right]$

$= \frac{1}{2}\frac{z}{z-e^{ia}} - \frac{1}{2}\frac{z}{z-e^{-ia}} = \frac{z}{2}\left(\frac{1}{z-e^{ia}} - \frac{1}{z-e^{-ia}}\right) = \frac{z}{2}\left(\frac{e^{ia}-e^{-ia}}{(z-e^{ia})(z-e^{-ia})}\right)$

$= z\left(\frac{\sin a}{(z-e^{ia})(z-e^{-ia})}\right) = \frac{z\sin a}{z^2 - z(e^{ia}+e^{-ia})+1} = \frac{z\sin a}{z^2 - 2z\cos a + 1}$

(21b) $\sin an \Leftrightarrow \dfrac{z\sin a}{z^2 - 2z\cos a + 1} = \dfrac{(\sin a)z^{-1}}{1-2(\cos a)z^{-1}+z^{-2}}$

Transform of $f(n)= b^n \cos(an)$

(22a) $f(n) = b^n \cos an$

$F(z) = Z[f(n)] = Z[b^n \cos an] = Z\left[\frac{1}{2}b^n e^{ian} + \frac{1}{2}b^n e^{-ian}\right] = Z\left[\frac{1}{2}(be^{ia})^n\right] + Z\left[\frac{1}{2}(be^{ia})^n\right]$

$= \frac{1}{2}\frac{z}{z-be^{ia}} + \frac{1}{2}\frac{z}{z-be^{-ia}} = \frac{z}{2}\left(\frac{1}{z-be^{ia}} + \frac{1}{z-be^{-ia}}\right)$

$= \frac{z}{2}\left(\frac{2z - be^{ia} - be^{-ia}}{(z-be^{ia})(z-be^{-ia})}\right) = z\left(\frac{z-b\cos a}{(z-be^{ia})(z-be^{-ia})}\right)$

$= z\dfrac{z-b\cos a}{z^2 - zb(e^{ia}+e^{-ia})+b^2} = z\dfrac{z-b\cos a}{z^2 - 2bz\cos a + b^2}$

(22b) $\cos an \Leftrightarrow z\dfrac{z-b\cos a}{z^2 - 2bz\cos a + b^2} = \dfrac{1-(b\cos a)z^{-1}}{1-(2b\cos a)z^{-1}+b^2 z^{-2}}$

Transform of $f(n)=b^n \sin(an)$

(23a) $f(n) = b^n \sin an$

$$F(z) = Z[f(n)] = Z[b^n \sin an] = Z\left[\tfrac{1}{2i}b^n e^{ian} - \tfrac{1}{2i}b^n e^{-ian}\right] = Z\left[\tfrac{1}{2i}(be^{ia})^n\right] - Z\left[\tfrac{1}{2i}(be^{-ia})^n\right]$$

$$= \frac{1}{2i}\frac{z}{z-be^{ia}} - \frac{1}{2i}\frac{z}{z-be^{-ia}} = \frac{z}{2i}\left(\frac{1}{z-be^{ia}} - \frac{1}{z-be^{-ia}}\right) = \frac{zb}{2i}\left(\frac{e^{ia}-e^{-ia}}{(z-be^{ia})(z-be^{-ia})}\right)$$

$$= zb\left(\frac{\sin a}{(z-be^{ia})(z-be^{-ia})}\right) = \frac{zb\sin a}{z^2 - zb(e^{ia}+e^{-ia})+b^2} = \frac{zb\sin a}{z^2 - 2zb\cos a + b^2}$$

(23b) $\sin an \Leftrightarrow \dfrac{zb\sin a}{z^2 - 2zb\cos a + b^2} = \dfrac{(b\sin a)z^{-1}}{1-(2b\cos a)z^{-1}+b^2 z^{-2}}$

11.4 Inverse Z Transforms

If we use the Z Transform to solve a problem in the z domain, then an *inverse* transform returns to the n domain.

A return from the *complex frequency domain* z to the n domain is achieved by performing the inverse operation. The operation is known as the Inverse Z Transform (equation 24).

(24) *inverse transform* $f(n) = \dfrac{1}{2\pi i}\oint F(z)z^{n-1}dz$

Figure 1102

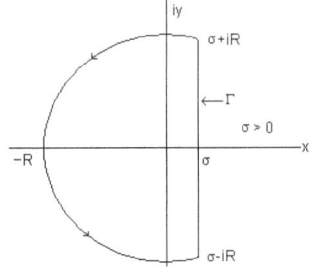

A straightforward method for integrating the inverse transform integral is to convert it to a closed contour Γ in the z plane so that the theory of residues can be used (for example Figure 1102).

(25) $f(n) = \dfrac{1}{2\pi i}\oint F(z)z^{n-1}dz = \sum residues$

The Γ contour requires all of the poles of F(z) to be enclosed by the contour.

The inverse operation using the inverse integral (24) is explained in Chapter 6 for the Laplace transform. The same process can be applied to the Z transform. However, as done with the Laplace transform, there is an easier way applicable to many problems that uses a partial fraction expansion (Section 4.5).

Mathematics beyond the Calculus

Example 1 The trick here is to divide F(z) by z.

(26a) $\dfrac{1}{z}F(z) = \dfrac{1}{z}\dfrac{5z-13}{(z-2)(z-3)} = -\dfrac{13}{6}\dfrac{1}{z} + \dfrac{3}{2}\dfrac{1}{(z-2)} + \dfrac{2}{3}\dfrac{1}{(z-3)}$

(26b) $F(z) = -\dfrac{13}{6} + \dfrac{3}{2}\dfrac{z}{(z-2)} + \dfrac{2}{3}\dfrac{z}{(z-3)}$

(26c) $f(n) = -\dfrac{13}{6}\delta(n) + \dfrac{3}{2}2^n u(n) + \dfrac{2}{3}3^n u(n)$ u(n) is unit step

Example 2

(27a) $\dfrac{1}{z}F(z) = \dfrac{1}{z}\dfrac{2z}{z^2-2z+2} = \dfrac{-i}{z-(1+i)} + \dfrac{i}{z-(1-i)}$

(27b) $F(z) = \dfrac{-iz}{z-(1+i)} + \dfrac{iz}{z-(1-i)}$

(27c) $f(n) = -i(1+i)^n u(n) + i(1-i)^n u(n)$ u(n) is unit step

(27d) $f(n) = \dfrac{-i^2}{i}(\sqrt{2}e^{i\pi/4})^n + \dfrac{i^2}{i}(\sqrt{2}e^{-i\pi/4})^n = 2(\sqrt{2})^n \left(\dfrac{e^{in\pi/4} - e^{-in\pi/4}}{2i} \right)$

$f(n) = 2(\sqrt{2})^n \sin\dfrac{\pi}{4}n$

Example 3

(28a) $F(z) = \dfrac{4z^2 + 5z^1 + 6}{z^2(z-1)} = \dfrac{4z^2 + 5z^1 + 6}{z^3} \dfrac{z}{(z-1)}$

(28b) $F(z) = (4z^{-1} + 5z^{-2} + 6z^{-3}) \times \dfrac{z}{(z-1)}$

(28c) $f(n) = (4\delta(n) + 5\delta(n-2) + 6\delta(n-3)) * u(n)$ u(n) is unit step

$f(n) = 4u(n) + 5u(n-2) + 6u(n-3)$

11 The Z Transform

Show that the transforms are correct when $Z[f(n)] = F(z)$

Problem 1101

$$\left(\frac{1}{2}\right)^n u(n) \Leftrightarrow \frac{1}{1-\frac{1}{2}z^{-1}} = \frac{z}{z-\frac{1}{2}} \quad |z| > \frac{1}{2}$$

Problem 1102

$$\left(\frac{1}{2}\right)^n u(-n) \Leftrightarrow \frac{-\frac{1}{2}z^{-1}}{1-\frac{1}{2}z^{-1}} = \frac{-\frac{1}{2}}{z-\frac{1}{2}} \quad |z| < \frac{1}{2}$$

Problem 1103

$$\left(\frac{1}{2}\right)^n \{u(n)-u(n-10)\} \Leftrightarrow \frac{1-\left(\frac{1}{2}z^{-1}\right)^{10}}{1-\frac{1}{2}z^{-1}} \quad |z| \neq 0$$

Problem 1105

$$f(n) = \begin{cases} n & 0 \leq n \leq N-1 \\ N & N \leq n \end{cases} \Leftrightarrow \frac{z^{-1}\left(1-z^{-N}\right)}{\left(1-z^{-1}\right)^2}$$

Problem 1106

$$-\frac{3}{2}\delta(n)+2u(n)-\frac{1}{2}2^n u(n) \Leftrightarrow \frac{z-3}{z^2-3z+2}$$

Problem 1107

$$-\frac{1}{6}\delta(n)-\frac{1}{2}2^n u(n)+\frac{2}{3}3^n u(n) \Leftrightarrow \frac{z-1}{(z-2)(z-3)}$$

Problem 1108

$$-\frac{1}{2}\delta(n)+\frac{5}{2}2^n u(n)+4n2^n u(n)+\frac{11}{2}n(n-1)2^{n-2} u(n) \Leftrightarrow \frac{2z^3+z^2-z+4}{(z-2)^3}$$

Problem 1109

$$2\delta(n)-9\left(\frac{1}{2}\right)^n u(n)+8u(n) \Leftrightarrow \frac{1+2z^{-1}+z^{-2}}{1-\frac{3}{2}z^{-1}+\frac{1}{2}z^{-2}}$$

Mathematics beyond the Calculus

General Transforms

(7b) $cf(n) \Leftrightarrow cF(z)$ (the $\Leftrightarrow$ symbol means transform from n to z or z to n)

(8b) $c_1 f_1(n) + c_2 f_2(n) \Leftrightarrow c_1 F_1(z) + c_2 F_2(z)$

(9b) $f(n-k)u(n-k) \Leftrightarrow z^{-k} F(z)$

(10b) $f(n-k) \Leftrightarrow z^{-k} \left[F(z) + \sum_{m=1}^{k} f(-m) z^m \right]$ form shows initial conditions

(11b) $nf(n) \Leftrightarrow -z \dfrac{dF(z)}{dz}$

(12b) $f_1(n) \times f_2(n) \Leftrightarrow F_1(z) F_2(z)$

(13b) $f(n+k) \Leftrightarrow z^{-k} \left[F(z) - \sum_{m=0}^{k-1} f(m) z^{-m} \right]$ this form shows initial conditions

(14) if $f(n) = 0$ for $n < 0$ then $\lim\limits_{z \to \infty} Z[f(n)] = f(0)$

Specific Transforms

(15b) $\delta(n-a) \Leftrightarrow z^{-a}$ $a > 0$ $\delta(n) \Leftrightarrow 1$

(16b) $a^n f(n) \Leftrightarrow F\left(\dfrac{z}{a}\right)$

(17b) $u(n) \Leftrightarrow \dfrac{z}{z-1} = \dfrac{1}{1-z^{-1}}$

(18b) $b^n u(n) \Leftrightarrow \dfrac{z}{z-b} = \dfrac{1}{1-bz^{-1}}$

(19b) $e^{an} u(n) \Leftrightarrow \dfrac{z}{z-e^a} = \dfrac{1}{1-e^a z^{-1}}$

(20b) $\cos an \Leftrightarrow z \dfrac{z - \cos a}{z^2 - 2z \cos a + 1} = \dfrac{1 - (\cos a) z^{-1}}{1 - 2(\cos a) z^{-1} + z^{-2}}$

(21b) $\sin an \Leftrightarrow \dfrac{z\sin a}{z^2-2z\cos a+1} = \dfrac{(\sin a)z^{-1}}{1-2(\cos a)z^{-1}+z^{-2}}$

(22b) $\cos an \Leftrightarrow z\dfrac{z-b\cos a}{z^2-2bz\cos a+b^2} = \dfrac{1-(b\cos a)z^{-1}}{1-(2b\cos a)z^{-1}+b^2 z^{-2}}$

(23b) $\sin an \Leftrightarrow \dfrac{zb\sin a}{z^2-2zb\cos a+b^2} = \dfrac{(b\sin a)z^{-1}}{1-(2b\cos a)z^{-1}+b^2 z^{-2}}$

$q^n \quad 0 < n \le N-1 \Leftrightarrow \dfrac{1-(qz^{-1})^N}{1-qz^{-1}}$

$nq^n u(n) \Leftrightarrow \dfrac{qz^{-1}}{(1-qz^{-1})^2} = \dfrac{qz}{(z-q)^2}$

$(n+1)q^n u(n) \Leftrightarrow \dfrac{1}{(1-qz^{-1})^2} = \dfrac{z^2}{(z-q)^2}$

$\dfrac{1}{2}(n+2)(n+1)q^n u(n) \Leftrightarrow \dfrac{1}{(1-qz^{-1})^3} = \dfrac{z^3}{(z-q)^3}$

$\dfrac{1}{2}n(n-1)a^n \Leftrightarrow \dfrac{a^2 z^{-2}}{(1-az^{-1})^3} = \dfrac{a^2 z}{(z-a)^3}$

Mathematics beyond the Calculus

12 Difference Equations

Difference equations are expressions representing relations of sequences of numbers x(n) *where n is an integer*. Some sequences of numbers x(n) are created by periodic sampling of continuous time signals where x(n) = f(nT) and T is the sampling period. Other sequences are derived from well known mathematical expressions such as the Fibonacci numbers. The sequences of numbers x(n) are analogous to functions such as f(y).

12.1 Elementary Sequences

The function $\delta(n)$ is analogous to the Dirac delta function (Figure 1101).

(1) $\delta(n) = \begin{cases} 1 & n = 0 \\ 0 & n \neq 0 \end{cases}$

Figure 1101 $\delta(n)$

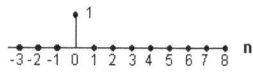

The sequence shown in Figure 1102 is written as y(n).

(2) $y(n) = a_{-3}\delta(n+3) + a_{-1}\delta(n+1) + a_1\delta(n-1) + a_3\delta(n-3) a_6\delta(n-6)$

Figure 1102 y(n)

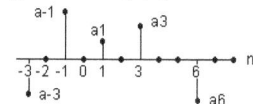

For example the unit step sequence u(n) is

(3) $u(n) = \begin{cases} 1 & n \geq 0 \\ 0 & n < 0 \end{cases}$

Figure 1103 u(n)

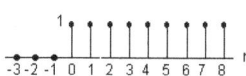

And then there are exponential sequences.

(4) $e(n) = \begin{cases} Aa^n & n \geq 0 \\ 0 & n < 0 \end{cases}$ (5) $e_1(n) = \begin{cases} e^{\frac{n}{2}} & n \geq 0 \\ 0 & n < 0 \end{cases}$

Figure 1104 $e_1(n)$

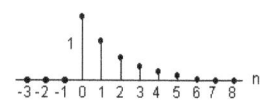

Here is a sinusoidal sequence.

(6) $x(n) = A\cos\omega n = A\cos\dfrac{2\pi n}{T}$

Figure 1105 x(n)

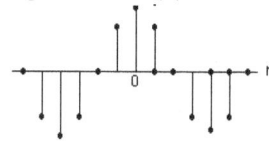

12 Difference Equations

12.2 Solution by Z Transform

The Z transform method is straightforward. The method for solving difference equations is implemented by the following steps.

1. Form the Z transform of both sides of the difference equation from the n domain to the z domain to produce the *subsidiary equation*, which is an algebraic equation with variable z. Manipulate the *subsidiary equation* to form Z[f(n)] = F(z) the ratio of two polynomials in z.
2. Expand F(z) into partial fractions. Invoke the initial conditions.
3. Perform the inverse transform to produce the solution $f(n) = Z^{-1}[F(z)]$.

Any sequence can be expressed as

(7) $x(n) = \sum_{j=-\infty}^{\infty} x(k)\delta(n-k)$

Z transforms facilitate the solutions of this type of equation.

(8a) Linear $Z[ax(n)+by(n)] = aZ[x(n)]+bZ[y(n)] = aX(z)+bY(z)$

(8b) Delay $Z[x(n-d)] = z^{-d} Z[x(n)] = z^{-d} X(z)$

(8c) Convolution $Z[x(n) \times y(n)] = X(z)Y(z)$

Example 1
(9a) $f(n)-3f(n-1)=0 \quad n \geq 0 \quad f(-1)=2 \quad Z[f(n)] = F(z)$

(9b) $F(z)-3z^{-1}[F(z)-zf(-1)] = 0$

$(1-3z^{-1})F(z) = 3f(-1) = 6$

$F(z) = \dfrac{6}{(1-3z^{-1})} = \dfrac{6z}{z-3}$

(9c) $f(n) = 6 \cdot 3^n u(n) \quad$ ref (18b) page 98

Example 2
(10a) $f(n)+4f(n-1)+4f(n-2)=0 \quad n \geq 0 \quad f(-1)=f(-2)=1$

(10b) $F(z)+4z^{-1}[F(z)+zf(-1)]+4z^{-2}[F(z)+zf(-1)++z^2 f(-2)] = 0$

$(1+4z^{-1}+4z^{-2})F(z) = -4f(-1)-4z^{-1}f(-1)-4f(-2) = -8-4z^{-1}$

$F(z) = \dfrac{-8-4z^{-1}}{1+4z^{-1}+4z^{-2}} = \dfrac{-8z^2-4z}{z^2+4z^1+4} = \dfrac{-8z^2-4z}{(z+2)^2} = \dfrac{-8z}{(z+2)} + \dfrac{12z}{(z+2)^2}$

(10c) $f(n) = -8(-2)^n u(n) - 6n(-2)^n u(n)$

Mathematics beyond the Calculus

Problem 1201
show that the solution to

$f(n+1)+f(n)=0$ where $f(-1)=1$ is $f(n)=(-1)^n$ $n>0$ hint - use (10b)

Problem 1202
show that the solution to

$f(n+2)+3f(n+1)+2f(n)=0$

where $f(-1)=0, f(-2)=1$ is $f(n)=(-1)^n-(-2)^n$ $n>0$ hint - use (10b)

Problem 1203
show that the solution to

$f(n+2)-3f(n+1)+2f(n)=-1$

where $f(-1)=2, f(-2)=4$ is $f(n)=2^n+n+1$ $n>0$ hint - use (10b)

Problem 1204
show that the solution to

$f(n+1)-2f(n)=n$

where $f(-1)=0$ is $f(n)=2^n-n-1$ $n>0$

Problem 1205
show that the solution to

$f(n+2)-f(n)=n+1$

where $f(-1)=0, f(-2)=0$ is $f(n)=\frac{1}{8}\left[(-1)^n+2n^2-1\right]$ $n>0$

Problem 1206
show that the solution to

$f(n+2)-\frac{3}{2}f(n+1)+\frac{1}{2}f(n)=\left(\frac{1}{3}\right)^n u(n)$ $f(0)=4, f(1)=0$

is $f(n)=-4\left(\frac{1}{2}\right)^n-1+9\left(\frac{1}{3}\right)^n$ $n>0$ hint - use (10b)

Problem 1207
show that the solution to

$f(n+3)+3f(n+2)-4f(n)=-n$ $\rightarrow$ $f(n)+3f(n-1)-4f(n-3)=-(n-3)$
where $f(-1)=f(-2)=f(-3)=0$ is $f(n)=\frac{1}{8}\left[(-1)^n+2n^2-1\right]$ $n>0$

13 Integral Equations

This is a hugh subject, which we limit to how transforms solve certain types of integral equations.

$$h(x)u(x) = f(x) + \int_a^{b(x)} K(x,t)u(t)dt \quad \text{where } u(t) \text{ is the unknown}$$

The type of equation is determined as follows

    $f(x) = 0$ Homogeneous
    $f(x) \neq 0$ Nonhomogeneous
    $b(x) = x$ Volterra integral equation
    $b(x) = b$ Fredholm integral equation
    $h(x) = 0$ Fredholm of the first kind
    $h(x) = 1$ Fredholm of the second kind

When the kernel $K(x,t)$ is in the form $K(x-t)$ the convolution integral results (equation 50 page 37).

$$I = \int_a^b K(x-t)u(t)dt$$

$$\mathcal{L}[I] = \mathcal{L}[K(x)] \times \mathcal{L}[u(t)]$$

Example 1

(1a) $h(x) = 1, \; K(x,t) = K(x-t) \;\rightarrow\; u(x) = f(x) + \lambda \int_0^x K(x-t)u(t)dt$

(1b) $U(p) = F(p) + \lambda K(p)U(p)$

$$U(p) = \frac{F(p)}{1 - \lambda K(p)}$$

$$u(x) = \mathcal{L}^{-1}\left\{\frac{F(p)}{1 - \lambda K(p)}\right\}$$

Example 2

(2a)   solve $u(x) = x - \int_0^x (t-x)u(t)dt$

(2b) $\mathcal{L}[u(x)] = \mathcal{L}[x] - \mathcal{L}[x]\mathcal{L}[u(x)] \;\rightarrow\; \mathcal{L}[u(x)] = \dfrac{1}{p^2} - \dfrac{1}{p^2}\mathcal{L}[u(x)]$

(2c) $\mathcal{L}[u(x)]\left(1 + \dfrac{1}{p^2}\right) = \dfrac{1}{p^2} \;\rightarrow\; \mathcal{L}[u(x)] = \dfrac{1}{p^2 + 1}$

(2d) $u(x) = \sin x$

Mathematics beyond the Calculus

Example 3

(3a) $\quad solve \quad u(x) = f(x) + \lambda \int_0^x e^{x-t} u(t) dt$

(3b) $\quad U(p) = F(p) + \lambda \mathcal{L}[e^x] \times U(p) = F(p) + \lambda \dfrac{1}{p-1} U(p)$

$$U(p) = \dfrac{F(p)}{1 - \lambda/(p-1)} = F(p) \dfrac{p-1-\lambda+\lambda}{p-1-\lambda} = F(p) + \lambda F(p) \dfrac{1}{p-(1+\lambda)}$$

(3b) $\quad u(x) = \mathcal{L}^{-1}[F(p)] + \lambda \mathcal{L}^{-1}\left[F(p) \cdot \dfrac{1}{p-(1+\lambda)}\right]$

$$u(x) = f(x) + \lambda \int_0^x e^{(1+\lambda)(x-t)} f(t) dt$$

Example 4

(4a) $\quad solve \quad \sin x = \int_0^x e^{x-t} u(t) dt$

(4b) $\quad \dfrac{1}{p^2+1} = \dfrac{1}{p-1} U(p) \rightarrow U(p) = \dfrac{p}{p^2+1} - \dfrac{1}{p^2+1}$

(4c) $\quad u(x) = \cos x - \sin x$

Example 5

(5a) $\quad solve \quad u(x) = 1 + \int_0^x u(t) dt = 1 + \int_0^x 1 \times u(t) dt$

(5b) $\quad \mathcal{L}[u(x)] = \mathcal{L}[1] + \mathcal{L}[1] \cdot \mathcal{L}[u(x)] \rightarrow \mathcal{L}[u(x)] = \dfrac{1}{p} + \dfrac{1}{p} \mathcal{L}[u(x)]$

(5c) $\quad \mathcal{L}[u(x)]\left(1 - \dfrac{1}{p}\right) = \dfrac{1}{p} \rightarrow \mathcal{L}[u(x)] = \dfrac{1}{p-1}$

(5d) $\quad u(x) = e^x$

Example 6

(6a) $\quad solve \quad \dfrac{\pi x}{2} = \int_0^x \dfrac{1}{\sqrt{x-t}} u(t) dt$

(6b) $\quad \mathcal{L}\left[\dfrac{\pi x}{2}\right] = \mathcal{L}\left[\dfrac{1}{\sqrt{x}}\right] \mathcal{L}[u(x)] \rightarrow \dfrac{\pi}{2p^2} = \dfrac{\sqrt{\pi}}{\sqrt{p}} \mathcal{L}[u(x)]$

(6c) $\quad \mathcal{L}[u(x)] = \dfrac{\sqrt{\pi}}{2p^{3/2}}$

(6d) $\quad u(x) = \sqrt{x}$

13 Integral Equations

Problem 1301

show that the solution to $x = \int_0^x \frac{1}{x-t} g(t) dt$ *is* $g(t) = 1 - t$

Problem 1302

show that the solution to $\sin x = \int_0^x J_0(x-t) g(t) dt$

where $J_0(x) \Leftrightarrow \dfrac{1}{(1+p^2)^{0.5}}$ *is* $g(t) = J_0(t)$ *and* $\sin x = \int_0^x J_0(x-t) J_0(t) dt$

Problem 1303

show that the solution to $f(x) = 1 - \int_0^x (x-t) g(t) dt$ *is* $g(t) = \cos t$

Problem 1304

show that the solution to $g(x) = 1 - \int_0^x (x-t) g(t) dt$ *is* $g(t) = \cos t$

Problem 1305

show that the solution to $g(x) = 1 + \int_0^x g(t) dt$ *is* $g(t) = e^t$

hint $\mathscr{L}\left[\int_0^t f(x) dx\right] = \dfrac{F(p)}{p}$

Problem 1306

show that the solution to $g(x) = \cos x - x - 2 + \int_0^x (t-x) g(t) dt$

is $g(t) = -\cos t - \sin t - \dfrac{t}{2} \sin t$

Problem 1307

show that the solution to $g(x) = 1 + 2\sin x - \int_0^x g(t) dt$

is $g(t) = \cos t + \sin t$

14 Galois Finite Fields GF(2^m)

Galois finite field GF(2^m) has 2^m elements where m is an integer. The field creates an arithmetic defining addition and multiplication for a finite number of 2^m elements. Galois finite fields are introduced, because their elements are used to implement error correcting code G and H matrices.

> Galois GF(2^m) adds to code theory, because α order equals the number of bits in a code word.
> $\alpha \text{ order} = 2^m - 1 \quad \text{bits in a code word} \quad n = 2^m - 1$

E. Galois' (1811-1832) finite field GF(2^m) *Constructing a Galois field* GF(2^m) starts with the 0 and 1 elements of GF(2) with modulus 2 arithmetic and a polynomial f(x) of degree m. The GF(2) arithmetic is

(1a) $0+0=0 \quad 0+1=1 \quad 1+0=1 \quad 1+1=0$
(1b) $0\times 0=0 \quad 0\times 1=0 \quad 1\times 0=0 \quad 1\times 1=1$

The symbol α is an element of GF(2^m). The order of α is 2^m-1.

(2) $\alpha^{2^m-1}=1$ and $0, 1, \alpha, \alpha^2, ..., \alpha^{2^m-2}$ are the elements of $GF(2^m)$

The new element α is defined as a root of some f(x) so that f(α)=0. If f(x) is selected properly the powers of α up to 2^m-2 are different, and they produce the elements of the Galois field.

For example let m=4. Then select what turns out to be a proper f(x) from the factors of $x^{2^m-1}+1=0$, which is derived from $\alpha^{2^m-1}=1$.
If $m=4$, then $x^{2^m-1}+1 = x^{2^4-1}+1 = x^{15}+1 = 0$ and so

(3) $x^{15}=1$ and the multiplicative order of any element is $\leq 2^4 - 1 = 15$

The elements of the field we want to construct are roots r_i of equation 3.

(4) $x^{2^m-1}+1 = x^{15}+1 = \prod_{i=1}^{15}(x-r_i)$

Divisors of 15 *are* 1, 3, 5, *and* 15, *use* 1
$x^{15}+1 = (x+1)(x^{14}+\cdots+1)$
Divisors of 14 *are* 2, 7, *use* 2
$x^{15}+1 = (x+1)(x^2+x+1)(x^{12}+....+1)$

14 Galois Finite Fields GF(2^m)

Divisors of 12 *are* 2, 3, 4, 6, *use* 4

(5) $x^{15} + 1 = (x+1)(x^2 + x + 1)(x^4 + x^3 + x^2 + x + 1)(x^4 + x^3 + 1)(x^4 + x + 1)$

We do not factor beyond equation 5, because the coefficients would not be 0 or 1.

(6) *select* $f(x) = x^4 + x + 1$ *a factor of* $x^{2^m - 1} + 1 = x^{15} + 1$

let $f(\alpha) = \alpha^4 + \alpha + 1 = 0$ [*define* α *as a root of* $f(x)$]

(7) $\alpha^4 = \alpha + 1$ (mod 2)

Alpha reduction For example consider constructing Galois field GF(2^4), which is an *extension field* of GF(2). Equation 7 provides the means whereby one can reduce any power of α to a polynomial of degree 3 or less in GF(2^4) as shown by the following algebraic manipulations, which confirm that the order of α is 15 (=$2^m - 1$, m=4). Order is the number of elements in the field.

0

$\alpha^0 = 1$

α^1

α^2

α^3

$\alpha^4 = \alpha + 1$

$\alpha^5 = \alpha^1 \alpha^4 = \alpha(\alpha + 1) = \alpha^2 + \alpha$

$\alpha^6 = \alpha^1 \alpha^5 = \alpha(\alpha^2 + \alpha) = \alpha^3 + \alpha^2$

$\alpha^7 = \alpha^1 \alpha^6 = \alpha(\alpha^3 + \alpha^2) = \alpha^3 + \alpha^4 = \alpha^3 + \alpha + 1$

$\alpha^8 = \alpha^1 \alpha^7 = \alpha(\alpha^3 + \alpha + 1) = \alpha^4 + \alpha^2 + \alpha = \alpha + 1 + \alpha^2 + \alpha = \alpha^2 + 1$

$\alpha^9 = \alpha^1 \alpha^8 = \alpha(\alpha^2 + 1) = \alpha^3 + \alpha$

$\alpha^{10} = \alpha^1 \alpha^9 = \alpha(\alpha^3 + \alpha) = \alpha^4 + \alpha^2 = \alpha^2 + \alpha + 1$

$\alpha^{11} = \alpha^1 \alpha^{10} = \alpha(\alpha^2 + \alpha + 1) = \alpha^3 + \alpha^2 + \alpha$

$\alpha^{12} = \alpha^1 \alpha^{11} = \alpha(\alpha^3 + \alpha^2 + \alpha) = \alpha^4 + \alpha^3 + \alpha^2 = \alpha^3 + \alpha^2 + \alpha + 1$

$\alpha^{13} = \alpha^1 \alpha^{12} = \alpha(\alpha^3 + \alpha^2 + \alpha + 1) = \alpha^4 + \alpha^3 + \alpha^2 + \alpha = \alpha^3 + \alpha^2 + 1$

$\alpha^{14} = \alpha^1 \alpha^{13} = \alpha(\alpha^3 + \alpha^2 + 1) = \alpha^4 + \alpha^3 + \alpha = \alpha + 1 + \alpha^3 + \alpha^2 = \alpha^3 + 1$

$\alpha^{15} = \alpha^1 \alpha^{14} = \alpha(\alpha^3 + 1) = \alpha^4 + \alpha = \alpha + 1 + \alpha = 1$

Mathematics beyond the Calculus

Recapitulation: Here are the elements of GF(16) created by x^4+x+1, their polynomials, and their 4-tuple binary number. Three representations of the GF(16) field elements are listed. Know that other representations are possible.

element	polynomial	4-tuple
$\alpha^{-\infty}$	0	0000
α^0	1	0001
α^1	α	0010
α^2	α^2	0100
α^3	α^3	1000
α^4	$\alpha+1$	0011
α^5	$\alpha^2+\alpha$	0110
α^6	$\alpha^3+\alpha^2$	1100
α^7	$\alpha^3+\alpha+1$	1011
α^8	α^2+1	0101
α^9	$\alpha^3+\alpha$	1010
α^{10}	$\alpha^2+\alpha+1$	0111
α^{11}	$\alpha^3+\alpha^2+\alpha$	1110
α^{12}	$\alpha^3+\alpha^2+\alpha+1$	1111
α^{13}	$\alpha^3+\alpha^2+1$	1101
α^{14}	α^3+1	1001
α^{15}	1	0001

Primitive element An element of GF(2^m), such as α, whose powers create 2^m distinct elements is a *primitive* element.

Primitive polynomial A polynomial of degree m, such as x^4+x+1 that creates 2^m distinct elements is a *primitive* polynomial.

The Arithmetic To multiply two elements simply add exponents and use the equation $\alpha^{15}=1$. Divide two elements by subtracting exponents.
$\alpha^8\alpha^{12} = \alpha^{20} = \alpha^{15}\alpha^5 = \alpha^5$

To add two elements add their polynomials modulo 2. For example
$\alpha^5 + \alpha^7 = (\alpha^2 + \alpha) + (\alpha^3 + \alpha + 1) = \alpha^3 + \alpha^2 + 1 = \alpha^{13}$

Addition is facilitated by Zech logarithms (Section 4.5 page 35) and the table above. For example
$\alpha^5 + \alpha^7 = \alpha^5(1+\alpha^2) = \alpha^5\alpha^8 = \alpha^{13}$

Problem 1401 Show that $f(x)=x^4+x^2+1$ is a primitive polynomial of GF(2^4).
Problem 1402 Construct GF(2^4) using non primitive $f(x)=x^4+x^3+x^2+x+1$.
Problem 1403 Show that α^4 is a primitive element of GF(2^4).

---

14 Galois Finite Fields GF(2^m)

Theorem 1 The number p of integral elements in the field is a prime number. If p=2, then the elements are 0 and 1.

Theorem 2 The number of elements in the field is a power of the prime p. *If p=2, then $p^m = 2^m$*

Theorem 3 Every element r_i of a field of order p^m satisfies the equation
$$x^{p^m-1} + x = \prod_{i=1}^{p^m-1}(x-r_i)$$
If p = 2, m = 3, then $x^{2^m-1}+1 = x^{2^3-1}+1 = x^7+1 = 0 \rightarrow x^7 = 1$
the multiplicative order of any element is $\leq p^m - 1$
and the additive order is p

Emphasis The additive order of an integral element is p, whereas the multiplicative order of any element is less than or equal to $p^m - 1$.

Theorem 4 If an equation of degree j belongs to the field, then it has at most j roots in the field.
$$x^7 + 1 = \prod_{i=0}^{7}(x-r_i)$$

Theorem 5 If d is a divisor of p^m-1, then the equation $x^d-1=0$ has d roots in a field of order p^m.

Definition An element of the field GF(2^m) having p^m-1 for its order is a primitive element of the field.

Theorem 8 All the elements of the field except the element 0 are powers of any given primitive element of the field. Hence the multiplicative group of the field is cyclic. For example

Order is $p^3 - 1 = 2^3 - 1 = 7$ in GF(8)
If α is a primative element, then $\alpha^7 = 1$ & all elements are powers of α
$\alpha^0, \alpha^1, \alpha^2, \alpha^3, \alpha^4, \alpha^5, \alpha^6$

Mathematics beyond the Calculus

14.1 Polynomial Operations

Polynomials in some variable x equivalent to binary n-tuples are a major addition to code structure. They are equivalent when coefficients of the powers of x are interpreted as the binary digits of a binary word (n-tuple).

The binary word 0100111 *may be represented by a polynomial*

(7a) $f(x) = 0x^6 + 1x^5 + 0x^4 + 0x^3 + 1x^2 + 1x^1 + 1x^0$

(7b) $f(x) = x^5 + x^2 + x + 1$

A polynomial of degree n−1 with binary coefficients has n terms each with coefficient 0 or 1. The set of n binary coefficients can represent one n bit code word, received word, error word, or columns or rows of bits in matrices.

The 2^k message words can be represented by 2^k polynomials of degree ≤ k with coefficients 0 and 1. Encoded messages with parity bits, code words, are represented by 2^n polynomials of degree n. And, 2^n polynomials represent the 2^n possible received words. A zero code word such as (0000000) corresponds to the f(x)=0 polynomial.

When we add, subtract, multiply, and divide polynomials, the operations on the binary coefficients are executed modulo 2 so that all resulting coefficients of x^j are 1 or 0.

Modular arithmetic limits the numbers that can appear in expressions to a range of numbers. The range is established in modular arithmetic by dividing all numbers by the modulus n and using only the n positive remainders 0, 1, ... , n−1. Modulo 2 remainders are 0 and 1 (page 23).

Addition and subtraction of the 0 and 1 coefficients are executed by modulo 2 arithmetic. The sum of any two polynomials is another polynomial of the same degree or less. Addition modulo 2 of coefficients implements 0+0=0, 0+1=1, 1+0=1, and 1+1=0. Subtraction operations produce the same results. THERE ARE NO CARRIES in any operation.

$$
\begin{array}{r}
x^5 + x^4 + x^3 + x^2 + x^1 \\
(8) \quad +x^5 \qquad\qquad +x^2 + x^1 + 1 \\ \hline
x^4 + x^3 \qquad\qquad +1
\end{array}
$$

14 Galois Finite Fields GF(2^m)

The degree of a product of two polynomials is the sum of the degrees of the two polynomials. For example here is a product of two polynomials executed in two formats.

(9)
$$\begin{array}{r}
x^3 + x^2 + 1 \\
\times\ x^3 + x^2 + x^1 \\
\hline
x^4 + x^3 + \quad\ x^1 \\
x^5 + x^4 \quad\ + x^2 \\
+\ x^6 + x^5 \quad\ + x^3 \\
\hline
x^6 \quad\quad\quad + x^2 + x^1
\end{array}
\qquad
\begin{array}{r}
1101 \\
\times 1110 \\
\hline
0000 \\
1101 \\
1101 \\
+1101 \\
\hline
1000110
\end{array}
\quad \rightarrow\ 1101 \times 1110 = 1000110$$

However the degree of a polynomial representing an n bit code word is limited to the range 0 to n−1. This means the degree of a product of polynomials has to be reduced modulo some polynomial, which is referred to as a *modulus polynomial*.

In modular arithmetic numbers are divided by the modulus and the remainder replaces the number divided. Then, by analogy, polynomials whose degree is greater than n−1 are divided by a modulus polynomial, and the remainder of degree n−1 (or less) replaces the polynomial divided. Consequently the remainder polynomial has n coefficients (or fewer), degree equal to n−1 (or less), and can represent an n-bit code word.

Multiplication of polynomials f(x) with degree f, and g(x) with degree g results in a polynomial h(x) with degree f+g. The binary coefficients of this polynomial cannot be interpreted as a binary n-tuple, because f+g > n. However, dividing h(x) by some irreducible p(x) with degree n produces remainder r(x) with degree n−1 or less. Then r(x) is considered to be the result of the original multiplication modulo p(x).

$$\frac{h(x)}{p(x)} = \frac{f(x)g(x)}{p(x)} = q(x) + \frac{r(x)}{p(x)}$$
$$h(x) = f(x)g(x) = q(x)p(x) + r(x) \quad [\deg r(x) < \deg p(x)]$$
(10) $\quad h(x) \equiv r(x)\ [\bmod p(x)]$
$$p(x)\ \text{divides}\ h(x) - r(x)$$

Problem 1404 Divide 0100000 by 1011 to get remainder 111.
Problem 1405 Divide 0110001 by 1011 to get remainder 000.
Problem 1406 Divide 1000000 by 1011 to get remainder 101.

Mathematics beyond the Calculus

P(x) must be irreducible (4.2). A specific division can be done three ways. We can use the polynomials directly, or we can replace the polynomials with binary numbers derived from the coefficients to simplify the writing of the division process. Or, we can use alpha reduction.

For example we calculate the remainder r(x) modulo p(x) of h(x) by long division and then compare the long division process to alpha reduction.

Long division Select $p(x)=x^4+x^3+1$ from the factors in equation 5 page 25, h(x) from equation 9 page 29.

(11) $p(x) \overline{)h(x)} \rightarrow x^4+x^3+1 \overline{)x^6+x^2+x} \rightarrow 11001 \overline{)1000110}$

```
             111
   11001 )1000110
           11001
           100010
            11001
             10000
             11001
              1001
```

$h(x) = q(x) \quad \times p(x) \quad + r(x)$
$x^6 + x^2 + x = (x^2 + x^1 + 1) \times (x^4 + x^3 + 1) + (x^3 + 1)$
$1000110 = 111 \times 11001 + 1001$
$1000110 = 1001 \pmod{11001}$
$h(x) = r(x) \bmod p(x)$
$r(x) = x^3 + 1$

Alpha reduction Long division is avoided when h(x) is changed to h(α), and *irreducible polynomial* $p(\alpha) = \alpha^4 + \alpha^3 + 1$ set to 0 defines $\alpha^4 = \alpha^3 + 1$. In effect $\alpha^4 = \alpha^3 + 1$ divides h(x) by $x^4 + x^3 + 1$.

(13) $h(\alpha) = \alpha^6 + \alpha^2 + \alpha = \alpha^4 \alpha^2 + \alpha^2 + \alpha = (\alpha^3 + 1)\alpha^2 + \alpha^2 + \alpha)$
$\quad = \alpha^5 + \alpha^2 + \alpha^2 + \alpha = \alpha^5 + \alpha = \alpha^4 \alpha + \alpha = (\alpha^3 + 1)\alpha + \alpha = \alpha^4 = \alpha^3 + 1$
$\quad = r(\alpha) \rightarrow r(x) = x^3 + 1$

> This procedure works for any h(x).

Problem 1407 Use $\alpha^4 = \alpha + 1$ to divide x^8 by $x^4 + x + 1$ to get $r(x) = x^2 + 1$.
Problem 1408 Use $\alpha^4 = \alpha + 1$ to divide x^6 by $x^4 + x + 1$ to get $r(x) = x^3 + x^2$.

14 Galois Finite Fields GF(2^m)

14.2 Irreducible Polynomials

An irreducible polynomial cannot be factored into factors with coefficients 0 and 1.

A polynomial f(x), of degree n, is *irreducible* over the binary field GF(2) if f(x) is not divisible by any polynomial of degree >0 and <n. One or more irreducible polynomials exist for every finite field GF(2^m).

The factors of $x^{2m-1}+1=0$ are irreducible polynomials.

Here is brute force checking of polynomials for divisibility.
Degree 0 $f(x) = 1$
Degree 1 $f(x) = x$ $f(x) = x+1$ *both are irreducible*
Degree 2 $f(x) = x^2 = x \cdot x$
$\qquad\qquad f(x) = x^2 + 1 = x^2 + 2x + 1 \pmod{2} = (x+1)^2$
$\qquad\qquad f(x) = x^2 + x = x(x+1)$
$\qquad\qquad f(x) = x^2 + x + 1 \quad (irreducible)$

The last polynomial is clearly irreducible when it is rewritten as
$f(x) = x^2 + x + 1 = x(x+1) + 1$ *so that*
$$\frac{f(x)}{x+1} = x + \frac{1}{x+1} \quad (remainder \neq 0 \ means \ f(x) \ is \ not \ divisible \ by \ x+1)$$

Degree 3 $\quad x^3 = x \cdot x \cdot x$
$\qquad\qquad x^3 + 1 = (x+1)(x^2 + x + 1)$
$\qquad\qquad x^3 + x = x(x^2 + 1) = x(x+1)^2$
$\qquad\qquad x^3 + x + 1 = x(x^2 + 1) + 1 \quad (irreducible)$
$\qquad\qquad x^3 + x^2 = x^2(x+1)$
$\qquad\qquad x^3 + x^2 + 1 = x^2(x+1) + 1 \quad (irreducible)$
$\qquad\qquad x^3 + x^2 + x = x(x^2 + x + 1)$
$\qquad\qquad x^3 + x^2 + x + 1 = x^2(x+1) + (x+1) = (x^2+1)(x+1) = (x+1)^3$

Problem 1409 Show that x^4+x+1 and $x^4+x^3+x^2+x+1$ are the only 4th degree irreducible polynomials.

Mathematics beyond the Calculus

14.3 Minimal Polynomials

We find the minimal polynomials m(x) of Galois field elements, which turn out to be irreducible and are the factors of $f(x) = x^{2^m} - x$.

> Minimal polynomials are used to form generator polynomials g(x), which in turn are used to produce the generator matrix G (4.6 p36).

Definition Let $\beta = \alpha^j$ be any element of Galois field GF(2^m). Then any polynomial m(x) of smallest degree with binary coefficients 0, 1 such that m(β)=0 is referred to as the *minimal polynomial* of β, which is also irreducible.

The degree of *minimal polynomials* equals the number of roots. One knows this from the fundamental theorem of algebra.

One can prove by induction that if f(x) is a polynomial in x belonging to GF(2^m) where all coefficients are 0 or 1, and if j is any positive integer, then in GF(2^m)

(14) $f(x^{2^j}) = [f(x)]^{2^j}$

This very unusual finite field property creates sets of roots of minimal polynomials for each power of α (Appendix A1 page 112). For example here is GF(2^4).

Table of sets of roots of $m_w(x)$ for GF(2^4) where $\alpha^{15} = 1$

	β^1	β^2	β^4	β^8	β^{16}	sets of roots
$\beta = \alpha^1$	α^1	α^2	α^4	α^8	$\alpha^{16} = \alpha^1$	$\alpha^1, \alpha^2, \alpha^4, \alpha^8$
$\beta = \alpha^2$	α^2	α^4	α^8	$\alpha^{16} = \alpha^1$	$\alpha^{32} = \alpha^2$	$\alpha^1, \alpha^2, \alpha^4, \alpha^8$
$\beta = \alpha^3$	α^3	α^6	α^{12}	$\alpha^{24} = \alpha^9$	$\alpha^{48} = \alpha^3$	$\alpha^3, \alpha^6, \alpha^9, \alpha^{12}$
$\beta = \alpha^4$	α^4	α^8	$\alpha^{16} = \alpha^1$	$\alpha^{32} = \alpha^2$	$\alpha^{64} = \alpha^4$	$\alpha^1, \alpha^2, \alpha^4, \alpha^8$
$\beta = \alpha^5$	α^5	α^{10}	$\alpha^{20} = \alpha^5$	$\alpha^{40} = \alpha^{10}$	$\alpha^{80} = \alpha^5$	α^5, α^{10}
$\beta = \alpha^6$	α^6	α^{12}	$\alpha^{24} = \alpha^9$	$\alpha^{48} = \alpha^3$	$\alpha^{96} = \alpha^6$	$\alpha^3, \alpha^6, \alpha^9, \alpha^{12}$
$\beta = \alpha^7$	α^7	α^{14}	$\alpha^{28} = \alpha^{13}$	$\alpha^{56} = \alpha^{11}$	$\alpha^{112} = \alpha^7$	$\alpha^7, \alpha^{11}, \alpha^{13}, \alpha^{14}$
:	:	:	:	:	:	:
$\beta = \alpha^{15}$	1	1	1	1	1	1
$\beta = \alpha^0$	1	1	1	1	1	1

14 Galois Finite Fields GF(2^m)

Each row in the Table lists the set of roots of a polynomial, which we name $m_w(x)$. Observe that the sets of roots in rows 1, 2, 4 are identical. This is why we can omit the $m_w(x)$ derived from β equal to even powers of α.

(15) *If* $m(β) = 0$, *then* $m(β^{2^j}) = [m(β)]^{2^j} = 0$

and $β^{2^0}, β^{2^1}, β^{2^2}, \cdots, β^{2^n}$ *are also roots of* $m(x)$

Observe in the following list that the subscript w of $m_w(x)$ is the power of the smallest root exponent. E. g. $m_0(x)$ has root $α^0$, $m_3(x)$ has root $α^3$, $m_5(x)$ has root $α^5$, etc.

For GF(2^4) construct polynomials $m_w(x)$ with roots shown in the Table.

$β = α^0 \to m_0(x) = (x - α^0) = x + 1$

$β = α^1 \to m_1(x) = (x - α^1)(x - α^2)(x - α^4)(x - α^8) = x^4 + x + 1$

$β = α^2 \to m_2(x) = m_1(x)$

$β = α^3 \to m_3(x) = (x - α^3)(x - α^6)(x - α^9)(x - α^{12}) = x^4 + x^3 + x^2 + x + 1$

$β = α^4 \to m_4(x) = m_1(x)$

$β = α^5 \to m_5(x) = (x - α^5)(x - α^{10}) = x^2 + x + 1$

$β = α^6 \to m_6(x) = m_3(x)$

$β = α^7 \to m_7(x) = (x - α^7)(x - α^{11})(x - α^{13})(x - α^{14}) = x^4 + x^3 + 1$

$\vdots$

$β = α^{15} \to m_{15}(x) = (x + 1)$

The odd numbered $m(x)$ include all of the roots. The even numbered $m(x)$ equal the odd numbered $m(x)$. Consequently only odd numbered $m_w(x)$ and $m_0(x)$ are included as factors of $x^{2^m-1} + 1 = 0$. For example

(16) $x^{2^4-1} + 1 = x^{15} + 1 = m_7(x)\, m_5(x)\, m_3(x)\, m_1(x)\, m_0(x)$

(17) $x^{15} + 1 = (x^4 + x^3 + 1)(x^2 + x + 1)(x^4 + x^3 + x^2 + x + 1)(x^4 + x + 1)(x + 1)$

Problem 1410 Reference equation 17. Multiply out the expression $m_0(x)\, m_1(x)\, m_3(x)\, m_5(x)\, m_7(x)$ to show that it equals $x^{15}+1$.

Mathematics beyond the Calculus

14.4 Primitive Roots and Primitive Polynomials

A *primitive* polynomial of degree m is an irreducible polynomial that is a factor of $x^{2^m-1}+1=0$ over $GF(2^m)$.

Definitions
1) A polynomial of degree m that produces a complete table of 2^m distinct elements (powers of α) of finite field $GF(2^m)$ is referred to as primitive.
2) If β is a primitive element of $GF(2^m)$, then an irreducible polynomial f(x) of degree m is primitive if $f(\beta)=0$ (4.3 page 32)

Primitives are important, because a primitive polynomial can serve as a generator of the finite set of elements of a finite field, and as a G matrix generator (Section 4.6 page 36).

All the elements of the field except the element α^0 are powers of any given primitive element of the field. Hence the multiplication group of the field is cyclic and finite.

(18) *If α is a primative element then $\alpha^{15}=1$ in GF(16) and all elements are powers of α. The distinct powers of α are*
$\alpha^0,\alpha^1,\alpha^2,\alpha^3,\alpha^4,\alpha^5,\alpha^6,\alpha^7,\alpha^8,\alpha^9,\alpha^{10},\alpha^{11},\alpha^{12},\alpha^{13},\alpha^{14}$

Examples Primitive polynomials are factors of $x^{2^m-1}+1=0$ in GF(4), GF(8), and GF(16).

(19a) $GF(4)$ $x^3+1=(x^2+x+1)(x+1)$
(19b) $GF(8)$ $x^7+1=(x^3+x^2+1)(x^3+x+1)(x+1)$
(19c) $GF(16)$ $x^{15}+1=(x^4+x^3+1)(x^2+x+1)(x^4+x^3+x^2+x+1)$
$(x^4+x+1)(x+1)$

The polynomial $x^{2^m-1}-1$ has as roots all of the 2^m-1 nonzero elements α^j of $GF(2^m)$

(2) $x^{2^m-1}+1=x^n+1=(x-\alpha^0)(x-\alpha^1)(x-\alpha^2)....(x-\alpha^{n-2})(x-\alpha^{n-1})$

14.5 Zech Logarithms

Zech logarithms are simply a convenient way to add powers of α. They are included here so that you know about them (there is an easier way). Zech logarithms are defined as follows.

(20) *Define the Zech logarithm as* $\alpha^{Z(n)} = \alpha^n + 1$

(21) $\alpha^p + \alpha^q = \alpha^q(\alpha^{p-q} + 1) = \alpha^q \alpha^{Z(p-q)} \rightarrow p - q = n$

Implementing addition this way requires a table of Z(n). The values of $\alpha^n + 1$ are found by using equations in the table of powers of α. Z(0) is not defined. Z(0) is not required, because the sum of two identical terms equals 0 (mod 2). Tables are found in Appendix A1 page 112.

Example GF(8) elements produce Z(n) for GF(8)

element	polynomial	3tuple			
α^0	1	001			$\alpha^{Z(n)} = \alpha^n + 1$
α^1	α	010	n		$\alpha^{Z(n)}$
α^2	α^2	100	0		$\alpha^0 + 1 = 0$
α^3	α + 1	011	1		$\alpha^1 + 1 = \alpha^3$
α^4	$\alpha^2 + \alpha$	110	2		$\alpha^2 + 1 = \alpha^6$
α^5	$\alpha^2 + \alpha + 1$	111	3		$\alpha^3 + 1 = \alpha$
α^6	$\alpha^2 + 1$	101	4		$\alpha^4 + 1 = \alpha^5$
α^7	1	001	5		$\alpha^5 + 1 = \alpha^4$
			6		$\alpha^6 + 1 = \alpha^2$

Example using Z(n) for GF(8) $\alpha^3 + \alpha^5 = \alpha^3(1 + \alpha^2) = \alpha^3 \alpha^6 = \alpha^9 = \alpha^2$
Using table of GF(8) elements $\alpha^3 + \alpha^5 = (\alpha + 1) + (\alpha^2 + \alpha + 1) = \alpha^2$

However there is an easier way. One method for *adding* powers of α involves representing them as polynomials in α, adding the binary coefficients of like terms modulo 2, and then converting the resulting sum polynomial back to a power of α.

> An easier way is to add the n-tuple representations.
> For example $\alpha^3 + \alpha^5 = 011 + 111 = 100 = \alpha^2$

14.6 Constructing a Systematic G Matrix Generator g(x)

Generator polynomial g(x) is a polynomial that produces the r parity bits' polynomial r(x) corresponding to the k message bits polynomial k(x).

Key consideration All of the factors of x^n-1 ($n=2^m-1$) are *minimal polynomials* $m_w(x)$ (Section 4.3 page 32). Let g(x) equal the product of one or more minimal polynomials $m_w(x)$. Which ones?

> The *generating polynomials* g(x) are products of one or more minimal polynomials $m_w(x)$ (Appendix A1 page 112), and the *degree of g(x) is r = n − k* as shown in equation 5 page 40.

Example let t=1, g(x) = $m_1(x)$ or $m_3(x)$
$GF(2^3)$, m=3, t=1, r=mt=3, $n=2^m-1=7$, k=n−r =4

(22a) $x^7 + 1 = m_3(x)m_1(x)m_0(x)$ $\qquad GF(2^3)$

(22b) $x^7 + 1 = (x^3 + x^2 + 1)(x^3 + x + 1)(x + 1)$

(22c) $r = 3 \rightarrow g_1(x) = m_1(x) = x^3 + x + 1$

Example let t=2, g(x) = $m_3(x) m_1(x)$
$GF(2^4)$, m=4, t=2, r=mt=8, $n=2^m-1=15$, k=n−r =7

(23a) $x^{15} + 1 = m_7(x)m_5(x)m_3(x)m_1(x)m_0(x)$ $\qquad GF(2^4)$

(23b) $x^{15} + 1 = (x^4 + x^3 + 1)(x^2 + x + 1)$
$\qquad\qquad (x^4 + x^3 + x^2 + x + 1)(x^4 + x + 1)(x + 1)$

(23c) $r = 8 \rightarrow g_2(x) = m_3(x)m_1(x) = x^8 + x^7 + x^6 + x^4 + 1$

Example let t=3, g(x) = $m_5(x) m_3(x) m_1(x)$
$GF(2^5)$, m=5, t=3, r=mt=15, $n=2^m-1=31$, k=n−r =16

(24a) $x^{31} + 1 = m_{15}(x)m_{11}(x)m_7(x)m_5(x)m_3(x)m_1(x)m_0(x)$

(24b) $x^{31} + 1 = (x+1)(x^5 + x^2 + 1)(x^5 + x^4 + x^3 + x^2 + 1)(x^5 + x^4 + x^2 + x^1 + 1)$
$\qquad\qquad (x^5 + x^3 + x^2 + x + 1)(x^5 + x^4 + x^3 + x + 1)(x^5 + x^3 + 1)$

(24c) $\qquad r = 15 \rightarrow g_3(x) = m_5(x)m_3(x)m_1(x)$
$\qquad\qquad = x^{15} + x^{11} + x^{10} + x^9 + x^8 + x^7 + x^5 + x^3 + x^2 + x + 1$

14.7 Systematic G and H Matrices are Related

We show by example that if G is known, then H is known, and vice versa. The basic reason that this is so is that each row of G is a code word C, and that the syndrome $S = H \times C^T = 0$.

Here is the table of $GF(2^3)$ elements.

element	polynomial		polynomial	3-tuple
α^0	1		1	001
α^1	α		α	010
α^2	α^2		α^2	100
α^3	$\alpha+1$		$\alpha+1$	011
α^4	$\alpha\alpha^3 = \alpha(\alpha+1)$		$\alpha^2+\alpha$	110
α^5	$\alpha^3\alpha^2 = (1+\alpha)\alpha^2 = \alpha^2+\alpha^3 = \alpha^2+\alpha+1$		$\alpha^2+\alpha+1$	111
α^6	$\alpha^3\alpha^3 = (\alpha+1)^2 = \alpha^2+0\alpha+1$		α^2+1	101
α^7	1		1	001

If we define the exponents of α as bit position numbers, then we rearrange the powers of α so that they are sequential.

(24) $H = \begin{bmatrix} \alpha^6 & \alpha^5 & \alpha^4 & \alpha^3 & \alpha^2 & \alpha^1 & \alpha^0 \end{bmatrix}$

We convert the alphas to binary n-tuples using the table above. The H matrix has two sub matrices P (parity) and I (unit).

(25) $H_{3\times 7} = \begin{bmatrix} 1 & 1 & 1 & 0 & 1 & 0 & 0 \\ 0 & 1 & 1 & 1 & 0 & 1 & 0 \\ 1 & 1 & 0 & 1 & 0 & 0 & 1 \end{bmatrix} = [P_{3\times 4} \mid I_{3\times 3}]$

The code word systematic assignments are as follows.

(26) $\begin{matrix} c_6 & c_5 & c_4 & c_3 & c_2 & c_1 & c_0 \\ m_3 & m_2 & m_1 & m_0 & r_2 & r_1 & r_0 \end{matrix}$

The syndrome reveals the three parity equations, because the syndrome of a no-error code word equals 0.

Mathematics beyond the Calculus

(27) $S_{3\times 1} = H_{3\times 7} C^T_{7\times 1} = \begin{bmatrix} m_3 + m_2 + m_1 + 0 + r_2 + 0 + 0 \\ 0 + m_2 + m_1 + m_0 + 0 + r_1 + 0 \\ m_3 + m_2 + 0 + m_0 + 0 + 0 + r_0 \end{bmatrix} = \begin{bmatrix} 0 \\ 0 \\ 0 \end{bmatrix}$

The corresponding encoder is the G matrix. G implements the parity equations. Watch the subscripts.

encoder $C = M \times G$

(28) $C = \begin{bmatrix} c_6 & c_5 & c_4 & c_3 & c_2 & c_1 & c_0 \end{bmatrix}$

$= \begin{bmatrix} m_3 & m_2 & m_1 & m_0 \end{bmatrix} \times \begin{bmatrix} I_{4\times 4} \mid P_{4\times 3} \end{bmatrix}$

$= \begin{bmatrix} m_3 & m_2 & m_1 & m_0 \end{bmatrix} \times \begin{bmatrix} 1 & 0 & 0 & 0 & 1 & 0 & 1 \\ 0 & 1 & 0 & 0 & 1 & 1 & 1 \\ 0 & 0 & 1 & 0 & 1 & 1 & 0 \\ 0 & 0 & 0 & 1 & 0 & 1 & 1 \end{bmatrix}$

Check:

$c_6 = m_3 + 0 + 0 + 0 = m_3$
$c_5 = 0 + m_2 + 0 + 0 = m_2$
$c_4 = 0 + 0 + m_1 + 0 = m_1$

(29) $c_3 = 0 + 0 + 0 + m_0 = m_0$
$c_2 = m_3 + m_2 + m_1 + 0 = r_2$
$c_1 = 0 + m_2 + m_1 + m_0 = r_1$
$c_0 = m_3 + m_2 + 0 + m_0 = r_0$

Note: For matrix transpose P^T see 1.4 page 8.

Conclusion for any n, k, r (k message elements, r parity elements, n=k+r)
Specific case: $P_{4\times 3}$ is the transpose of $P_{3\times 4}$ → $P_{4\times 3} = P_{3\times 4}^T$
General case: $P_{k\times r}$ is the transpose of $P_{r\times k}$ → $P_{k\times r} = P_{r\times k}^T$

(30a) if $H = \begin{bmatrix} P_{r\times k} \mid I_{r\times r} \end{bmatrix}$ then $G = \begin{bmatrix} I_{k\times k} \mid P_{k\times r} \end{bmatrix} = \begin{bmatrix} I_{k\times k} \mid P_{r\times k}^T \end{bmatrix}$

(30b) if $G = \begin{bmatrix} I_{k\times k} \mid P_{k\times r} \end{bmatrix}$ then $H = \begin{bmatrix} P_{r\times k} \mid I_{r\times r} \end{bmatrix} = \begin{bmatrix} P_{k\times r}^T \mid I_{r\times r} \end{bmatrix}$

Observe the different size of the I matrices in equations 30a and 30b.

Problem 1411 Repeat the above process creating G and H for GF(16).

Appendix

A1 Galois Field Equations

Finite field elements α^n, minimal polynomials m(x), generating polynomials g(x) for t errors, and Zech logarithms Z(n) are listed for finite fields $GF(2^3)$, $GF(2^4)$, $GF(2^5)$, and $GF(2^6)$.

Finite fields can be constructed when the number of elements u is an integer power m of any prime number q such as 2. We are only interested in binary numbers so $u=q^m=2^m$. We construct finite field $GF(2^m)$, which is an *extension field* of GF(2).

The key equation is $x^{2^m-1}+1=0$.

This equation is satisfied if any of the *factors* of this polynomial are equal to zero. All factors are irreducible.

Any polynomial that is a factor of $x^{2^m-1}+1=0$ is a *primitive polynomial*. A primitive polynomial always exists, and thus there will always be a primitive element α

Note: The order of α is 2^m-1. Order is the number of elements.

The degree of the primitive polynomials for $GF(2^m)$ is always m.

The element 0 is written as $\alpha^{-\infty}$

The element 1 is written as α^0

Define the Zech logarithm as $\alpha^{Z(n)} = \alpha^n + 1$
$\alpha^n + \alpha^m = \alpha^m(\alpha^{n-m}+1) = \alpha^m \alpha^{Z(n-m)} = \alpha^{m+Z(n-m)}$

A1.1 GF(2^3)

If $q = 2$, $m = 3$, then $x^{q^m-1} - 1 = x^{2^3-1} - 1 = x^7 - 1 = 0$

$x^7 = 1$

$x^{2^m-1} - 1 = x^7 - 1 = \prod_{i=1}^{7}(x - u_i)$

Divisors of $n = 7$ are 1 and 7

$x^7 - 1 = (x+1)(x^6 + \cdots + 1)$

factor x^6 Divisors of 6 are 1, 2, 3, and 6

$x^7 - 1 = (x+1)(x^3 + x + 1)(x^3 + x^2 + 1)$ (mod 2)

where all factors are irreducible

If $f(x) = x^3 + x + 1$, then $f(\alpha) = \alpha^3 + \alpha + 1 = 0$

$\alpha^3 = \alpha + 1$ define α to be a root of $f(x)$

order is $n = q^3 - 1 = 2^3 - 1 = 7$ the number of elements

If α is a primative element then $\alpha^7 = 1$ and all elements are powers of α

α^0, α^1, α^2, α^3, α^4, α^5, α^6, α^7

$\alpha^0 = 1$
$\alpha^1 = \alpha$
$\alpha^2 = \alpha^2$
$\alpha^3 = \alpha + 1$
$\alpha^4 = \alpha \alpha^3 = \alpha(\alpha + 1) = \alpha^2 + \alpha$
$\alpha^5 = \alpha^2 \alpha^3 = \alpha^2(\alpha + 1) = \alpha^3 + \alpha^2 = \alpha^2 + \alpha + 1$
$\alpha^6 = \alpha^3 \alpha^3 = (\alpha + 1)(\alpha + 1) = \alpha^2 + 2\alpha + 1 = \alpha^2 + 1$
$\alpha^7 = \alpha \alpha^6 = \alpha(\alpha^2 + 1) = \alpha^3 + \alpha = \alpha + 1 + \alpha = 1$

The order of α is $n = 7$ $\rightarrow$ $\alpha^7 = 1$

Use the multiplication property of the field to form powers of n-tuples

$\alpha^5 \rightarrow 111$

$(\alpha^5)^3 = \alpha^{15} = \alpha^{2\times 7+1} = \alpha^1 \rightarrow 010$

$(111)^3 = 010$

Appendix

GF(8) - the $m_i(x)$
$$x^7 - 1 = (x+1)(x^3 + x + 1)(x^3 + x^2 + 1)$$

factor	roots	minimal polynomial	
$x+1$	α^0, α^7	$m_0(x)$	11
$x^3 + x + 1$	$\alpha^1, \alpha^2, \alpha^4$	$m_1(x)$	1011
$x^3 + x^2 + 1$	$\alpha^3, \alpha^5, \alpha^6$	$m_3(x)$	1101

GF(8) - the $g_i(x)$ r=mt=3t

t	g(x)	binary	octal	n	k	r
1	$m_1(x)$	1011	13	7	4	3
2	$m_3(x)m_1(x)$	1111111	177	7	1	6

GF(8) - the elements

element	polynomial	3-tuple
α^0	1	001
α^1	α	010
α^2	α^2	100
α^3	$\alpha + 1$	011
α^4	$\alpha^2 + \alpha$	110
α^5	$\alpha^2 + \alpha + 1$	111
α^6	$\alpha^2 + 1$	101
α^7	1	001

GF(8) Z(n) Zech Logarithms
$\alpha^{Z(n)} = \alpha^n + 1$

n	$\alpha^{Z(n)}$	Z(n)
0	$\alpha^0 + 1 = 1 + 1 = 0$	$-\infty$
1	$\alpha^1 + 1 = \alpha^3$	3
2	$\alpha^2 + 1 = \alpha^6$	6
3	$\alpha^3 + 1 = \alpha + 1 + 1 = \alpha$	1
4	$\alpha^4 + 1 = \alpha^2 + \alpha + 1 = \alpha^5$	5
5	$\alpha^5 + 1 = \alpha^2 + \alpha + 1 + 1 = \alpha^4$	4
6	$\alpha^6 + 1 = \alpha^2 + 1 + 1 = \alpha^2$	2

Multiplication – add exponents modulo 7.
For example $\alpha^{15} \alpha^8 = \alpha^{23} = \alpha^{21}\alpha^2 = \alpha^7 \alpha^7 \alpha^7 \alpha^2 = \alpha^0 \alpha^0 \alpha^0 \alpha^2 = \alpha^2$

Addition Table GF(2^3)

	α^0	α^1	α^2	α^3	α^4	α^5	α^6	α^7
α^0	$\alpha^{-\infty}$	α^3	α^6	α^1	α^5	α^4	α^2	$\alpha^{-\infty}$
α^1		$\alpha^{-\infty}$	α^4	α^0	α^2	α^6	α^5	α^3
α^2			$\alpha^{-\infty}$	α^5	α^1	α^3	α^0	α^6
α^3				$\alpha^{-\infty}$	α^6	α^2	α^4	α^1
α^4					$\alpha^{-\infty}$	α^0	α^3	α^5
α^5						$\alpha^{-\infty}$	α^1	α^4
α^6							$\alpha^{-\infty}$	α^2
α^7								$\alpha^{-\infty}$

A1.2 GF(2^4)

If $q = 2$, $m = 4$, then $x^{q^m-1} - 1 = x^{2^4-1} - 1 = x^{15} - 1 = 0$

$x^{15} = 1$

$x^{2^m-1} - 1 = x^{15} - 1 = \prod_{i=1}^{15}(x - u_i)$

Divisors of $n = 15$ are 1, 3, 5, and 15

$x^{15} - 1 = (x+1)(x^{14} + \cdots + 1)$

factor x^{14}... Divisors of 14 are 2, 7

$x^{15} - 1 = (x+1)(x^2 + x + 1)(x^{12} + \cdots + 1)$ (mod 2)

factor x^{12} Divisors of 12 are 2, 3, 4, 6

$x^{15} - 1 = (x+1)(x^2 + x + 1)(x^4 + x^3 + x^2 + x + 1)(x^4 + x^3 + 1)(x^4 + x + 1)$

where all factors are irreducible

If $f(x) = x^4 + x + 1$, then $f(\alpha) = \alpha^4 + \alpha + 1 = 0$ [α is a root of $f(x)$]

$\alpha^4 = \alpha + 1$ (mod 2)

order is $n = 2^4 - 1 = 2^4 - 1 = 15$

If α is a primative element then $\alpha^{15} = 1$ & all elements are powers of α

$\alpha^0, \alpha^1, \alpha^2, \alpha^3, \alpha^4, \alpha^5, \alpha^6, \alpha^7, \alpha^8, \alpha^9, \alpha^{10}, \alpha^{11}, \alpha^{12}, \alpha^{13}, \alpha^{14}, \alpha^{15}$

$\alpha^0 \quad \alpha^1 \quad \alpha^2 \quad \alpha^3$

$\alpha^4 = \alpha + 1$

$\alpha^5 = \alpha^1 \alpha^4 = \alpha(\alpha + 1) = \alpha^2 + \alpha$

$\alpha^6 = \alpha^1 \alpha^5 = \alpha(\alpha^2 + \alpha) = \alpha^3 + \alpha^2$

$\alpha^7 = \alpha^1 \alpha^6 = \alpha(\alpha^3 + \alpha^2) = \alpha^4 + \alpha^3 = \alpha^3 + \alpha + 1$

$\alpha^8 = \alpha^1 \alpha^7 = \alpha(\alpha^3 + \alpha + 1) = \alpha^4 + \alpha^2 + \alpha = \alpha + 1 + \alpha^2 + \alpha = \alpha^2 + 1$

$\alpha^9 = \alpha^1 \alpha^8 = \alpha(\alpha^2 + 1) = \alpha^3 + \alpha$

$\alpha^{10} = \alpha^1 \alpha^9 = \alpha(\alpha^3 + \alpha) = \alpha^4 + \alpha^2 = \alpha^2 + \alpha + 1$

$\alpha^{11} = \alpha^1 \alpha^{10} = \alpha(\alpha^2 + \alpha + 1) = \alpha^3 + \alpha^2 + \alpha$

$\alpha^{12} = \alpha^1 \alpha^{11} = \alpha(\alpha^3 + \alpha^2 + \alpha) = \alpha^4 + \alpha^3 + \alpha^2 = \alpha^3 + \alpha^2 + \alpha + 1$

$\alpha^{13} = \alpha^1 \alpha^{12} = \alpha(\alpha^3 + \alpha^2 + \alpha + 1) = \alpha^4 + \alpha^3 + \alpha^2 + \alpha = \alpha^3 + \alpha^2 + 1$

$\alpha^{14} = \alpha^1 \alpha^{13} = \alpha(\alpha^3 + \alpha^2 + 1) = \alpha^4 + \alpha^3 + \alpha = \alpha + 1 + \alpha^3 + \alpha^2 = \alpha^3 + 1$

$\alpha^{15} = \alpha^1 \alpha^{14} = \alpha(\alpha^3 + 1) = \alpha^4 + \alpha = \alpha + 1 + \alpha = 1$

Appendix

The order of α is $n = 15$ $\rightarrow$ $\alpha^{15} = 1$
Use the multiplication property of the field to form powers of n-tuples
$\alpha^6 \rightarrow 1100$
$(\alpha^6)^7 = \alpha^{42} = \alpha^{2 \times 15 + 12} = \alpha^{12} \rightarrow 1111$
$(1100)^7 = 1111$

GF(16) - the $m_i(x)$

$x^{15} + 1 = m_0(x) m_1(x) m_3(x) m_5(x) m_7(x)$

factor	roots	minimal polynomial	
$x+1$	α^0	$m_0(x)$	00011
$x^4 + x + 1$	$\alpha^1, \alpha^2, \alpha^4, \alpha^8$	$m_1(x)$	10011
$x^4 + x^3 + x^2 + x + 1$	$\alpha^3, \alpha^6, \alpha^{12}, \alpha^9$	$m_3(x)$	11111
$x^2 + x + 1$	α^5, α^{10}	$m_5(x)$	00111
$x^4 + x^3 + 1$	$\alpha^7, \alpha^{11}, \alpha^{13}, \alpha^{14}$	$m_7(x)$	11001

GF(16) - the $g_i(x)$

t	g(x)	binary	octal	n	k	r
1	$m_1(x)$	10 011	23	15	11	4
2	$m_3(x) m_1(x)$	111 010 001	721	15	7	8
3	$m_7(x) m_3(x) m_1(x)$	1001 001 001 001	11111	15	3	12

GF(16) elements

element	polynomial	4tuple	hex	element	polynomial	binary	hex
$\alpha^{-\infty}$	0	0000	0	α^7	$\alpha^3 + \alpha^1 + \alpha^0$	1011	B
α^0	1	0001	1	α^8	$\alpha^2 + \alpha^0$	0101	5
α^1	α^1	0010	2	α^9	$\alpha^3 + \alpha^1$	1010	A
α^2	α^2	0100	4	α^{10}	$\alpha^2 + \alpha^1 + \alpha^0$	0111	7
α^3	α^3	1000	8	α^{11}	$\alpha^3 + \alpha^2 + \alpha^1$	1110	E
α^4	$\alpha^1 + \alpha^0$	0011	3	α^{12}	$\alpha^3 + \alpha^2 + \alpha^1 + \alpha^0$	1111	F
α^5	$\alpha^2 + \alpha^1$	0110	6	α^{13}	$\alpha^3 + \alpha^2 + \alpha^0$	1101	D
α^6	$\alpha^3 + \alpha^2$	1100	C	α^{14}	$\alpha^3 + \alpha^0$	1001	9
				α^{15}	α^0	0001	1

Mathematics beyond the Calculus

GF(16) Z(n) $\qquad \alpha^{z(n)} = \alpha^n + 1$

n	$\alpha^{Z(n)}$	$Z(n)$
0	$\alpha^0 + 1 = 1 + 1 = 0$	$-\infty$
1	$\alpha^1 + 1 = \alpha^4$	4
2	$\alpha^2 + 1 = \alpha^8$	8
3	$\alpha^3 + 1 = \alpha^{14}$	14
4	$\alpha^4 + 1 = \alpha + 1 + 1 = \alpha^1$	1
5	$\alpha^5 + 1 = \alpha^2 + \alpha + 1 = \alpha^{10}$	10
6	$\alpha^6 + 1 = \alpha^3 + \alpha^2 + 1 = \alpha^{13}$	13
7	$\alpha^7 + 1 = \alpha^3 + \alpha + 1 + 1 = \alpha^9$	9
8	$\alpha^8 + 1 = \alpha^2 + 1 + 1 = \alpha^2$	2
9	$\alpha^9 + 1 = \alpha^3 + \alpha + 1 = \alpha^7$	7
10	$\alpha^{10} + 1 = \alpha^2 + \alpha + 1 + 1 = \alpha^5$	5
11	$\alpha^{11} + 1 = \alpha^3 + \alpha^2 + \alpha + 1 = \alpha^{12}$	12
12	$\alpha^{12} + 1 = \alpha^3 + \alpha^2 + \alpha + 1 + 1 = \alpha^{11}$	11
13	$\alpha^{13} + 1 = \alpha^3 + \alpha^2 + 1 + 1 = \alpha^6$	6
14	$\alpha^{14} + 1 = \alpha^3 + 1 + 1 = \alpha^3$	3

Multiplication – add exponents modulo 15.
For example $\alpha^{27} \alpha^8 = \alpha^{35} = \alpha^{30} \alpha^5 = \alpha^{15} \alpha^{15} \alpha^5 = \alpha^0 \alpha^0 \alpha^5 = \alpha^5$

Addition Table GF(2^4) $\quad \gamma = \alpha^{-\infty}$ Only exponents are shown in this table.

	0	1	2	3	4	5	6	7	8	9	10	11	12	13	14
0	γ	4	8	14	1	10	13	9	2	7	5	12	11	6	3
1		γ	5	9	0	2	11	14	10	3	8	6	13	12	7
2			γ	6	10	1	3	12	0	11	4	9	7	14	13
3				γ	7	11	2	4	13	1	12	5	10	8	0
4					γ	8	12	3	5	14	2	13	6	11	9
5						γ	9	13	4	6	0	3	14	7	12
6							γ	10	14	5	7	1	4	0	8
7								γ	11	0	6	8	2	5	1
8									γ	12	1	7	9	3	6
9										γ	13	2	8	10	4
10											γ	14	3	9	11
11												γ	0	4	10
12													γ	1	5
13														γ	2
14															γ

A1.3 GF(2^5)

If $q = 2$, $m = 5$, then $x^{q^m-1} - 1 = x^{2^5-1} - 1 = x^{31} - 1 = 0$

$x^{31} = 1$

$x^{2^m-1} - 1 = x^{31} - 1 = \prod_{i=1}^{31}(x - u_i)$

Divisors of $n = 31$ are $1, 31$

$x^{31} - 1 = (x+1)(x^{30} + \cdots + 1)$

factor x^{30}....Divisors of 30 are 2, 3, 5

$x^{31} - 1 = (x+1)(x^5 + x^2 + 1)(x^{25} + \cdots + 1)$ (mod 2)

factor x^{25} Divisors of 25 are 5

$x^{31} - 1 = (x+1)(x^5 + x^2 + 1)(x^5 + x^4 + x^3 + x^2 + 1)(x^5 + x^4 + x^2 + x^1 + 1)$
$(x^5 + x^3 + x^2 + x + 1)(x^5 + x^4 + x^3 + x + 1)(x^5 + x^3 + 1)$ (mod 2)

If $f(x) = x^5 + x^2 + 1$, then $f(\alpha) = \alpha^5 + \alpha^2 + 1 = 0$ [α is a root of $f(x)$]

$\alpha^5 = \alpha^2 + 1$ (mod 2)

order is $n = 2^5 - 1 = 2^5 - 1 = 31$

If α is a primative element then $\alpha^{31} = 1$ & all elements are powers of α

$\alpha^0, \alpha^1, \alpha^2, \alpha^3, \alpha^4, \alpha^5, \alpha^6, \alpha^7, \alpha^8, \alpha^9, \alpha^{10}, \alpha^{11}, \alpha^{12}, \alpha^{13}, \alpha^{14}, \alpha^{15}, \ldots, \alpha^{30}, \alpha^{31}$

The order of α is $n = 31$ → $\alpha^{31} = 1$

Use the multiplication property of the field to form powers of n-tuples

$\alpha^6 \to 01010$

$(\alpha^6)^7 = \alpha^{42} = \alpha^{31+11} = \alpha^{11} \to 00111$

$(01010)^7 = 00111$

GF(32) - the $m_i(x)$

$x^{31} - 1 = (x+1)(x^5 + x^2 + 1)(x^5 + x^4 + x^3 + x^2 + 1)(x^5 + x^4 + x^2 + x^1 + 1)$
$(x^5 + x^3 + x^2 + x + 1)(x^5 + x^4 + x^3 + x + 1)(x^5 + x^3 + 1)$

Mathematics beyond the Calculus

factor	roots	minimal polynomial
$x+1$	α^0	$m_0(x)$ 000011
x^5+x^2+1	$\alpha^1,\alpha^2,\alpha^4,\alpha^8,\alpha^{16}$	$m_1(x)$ 100101
$x^5+x^4+x^3+x^2+1$	$\alpha^3,\alpha^6,\alpha^{12},\alpha^{24},\alpha^{17}$	$m_3(x)$ 111101
$x^5+x^4+x^2+x^1+1$	$\alpha^5,\alpha^{10},\alpha^{20},\alpha^9,\alpha^{18}$	$m_5(x)=m_9(x)$ 110111
$x^5+x^3+x^2+x+1$	$\alpha^7,\alpha^{14},\alpha^{28},\alpha^{25},\alpha^{19}$	$m_7(x)$ 101111
$x^5+x^4+x^3+x+1$	$\alpha^{11},\alpha^{22},\alpha^{13},\alpha^{26},\alpha^{21}$	$m_{11}(x)$ 111011
x^5+x^3+1	$\alpha^{15},\alpha^{30},\alpha^{29},\alpha^{27},\alpha^{23}$	$m_{15}(x)$ 101001

GF(32) $g_i(x)$ (most binaries not shown)

t	$g(x)$	binary	octal	n	k	$r=5t$
1	$m_1(x)$	100101	45	31	26	5
2	$m_3(x)m_1(x)$	–	3551	31	21	10
3	$m_5(x)m_3(x)m_1(x)$	–	107657	31	16	15
4 or 5	$m_7(x)m_5(x)m_3(x)m_1(x)$	–	5423325	31	11	20
6 or 7	$m_{11}(x)m_7(x)m_5(x)m_3(x)m_1(x)$	–	313365047	31	6	25
8 to 15	$m_{15}(x)m_{11}(x)m_7(x)m_5(x)m_3(x)m_1(x)$	–	17777777777	31	1	30

Appendix

GF(32) elements

element	polynomial	5tuple
$\alpha^{-\infty}$	0	00000
α^0	1	00001
α^1	α^1	00010
α^2	α^2	00100
α^3	α^3	01000
α^4	α^4	10000
α^5	$\alpha^2 + \alpha^0$	00101
α^6	$\alpha^3 + \alpha^1$	01010
α^7	$\alpha^4 + \alpha^2$	10100
α^8	$\alpha^3 + \alpha^2 + \alpha^0$	01101
α^9	$\alpha^4 + \alpha^3 + \alpha^1$	11010
α^{10}	$\alpha^4 + \alpha^0$	10001
α^{11}	$\alpha^2 + \alpha^1 + \alpha^0$	00111
α^{12}	$\alpha^3 + \alpha^2 + \alpha^1$	01110
α^{13}	$\alpha^4 + \alpha^3 + \alpha^2$	11100
α^{14}	$\alpha^4 + \alpha^3 + \alpha^2 + \alpha^0$	11101
α^{15}	$\alpha^4 + \alpha^3 + \alpha^2 + \alpha^1 + \alpha^0$	11111
α^{16}	$\alpha^4 + \alpha^3 + \alpha^1 + \alpha^0$	11011
α^{17}	$\alpha^4 + \alpha^1 + \alpha^0$	10011
α^{18}	$\alpha^1 + \alpha^0$	00011
α^{19}	$\alpha^2 + \alpha^1$	00110
α^{20}	$\alpha^3 + \alpha^2$	01100
α^{21}	$\alpha^4 + \alpha^3$	11000
α^{22}	$\alpha^4 + \alpha^2 + \alpha^0$	10101
α^{23}	$\alpha^3 + \alpha^2 + \alpha^1 + \alpha^0$	01111
α^{24}	$\alpha^4 + \alpha^3 + \alpha^2 + \alpha^1$	11110
α^{25}	$\alpha^4 + \alpha^3 + \alpha^0$	11001
α^{26}	$\alpha^4 + \alpha^2 + \alpha^1 + \alpha^0$	10111
α^{27}	$\alpha^3 + \alpha^1 + \alpha^0$	01011
α^{28}	$\alpha^4 + \alpha^2 + \alpha^1$	10110
α^{29}	$\alpha^3 + \alpha^0$	01001
α^{30}	$\alpha^4 + \alpha^1$	10010
α^{31}	α^0	00001

Mathematics beyond the Calculus

GF(32) Z(n)

$$\alpha^{Z(n)} = \alpha^n + 1$$

n	$\alpha^{Z(n)}$	$Z(n)$
0	$\alpha^0 + 1 = 1 + 1 = 0$	$-\infty$
1	$\alpha^1 + 1 = \alpha^{18}$	18
2	$\alpha^2 + 1 = \alpha^5$	5
3	$\alpha^3 + 1 = \alpha^{29}$	29
4	$\alpha^4 + 1 = \alpha^{10}$	10
5	$\alpha^5 + 1 = \alpha^2 + 1 + 1 = \alpha^2$	2
6	$\alpha^6 + 1 = \alpha^3 + \alpha + 1 = \alpha^{27}$	27
7	$\alpha^7 + 1 = \alpha^4 + \alpha^2 + 1 = \alpha^{22}$	22
8	$\alpha^8 + 1 = \alpha^3 + \alpha^2 + 1 + 1 = \alpha^{20}$	20
9	$\alpha^9 + 1 = \alpha^4 + \alpha^3 + \alpha^1 + 1 = \alpha^{16}$	16
10	$\alpha^{10} + 1 = \alpha^4 + 1 + 1 = \alpha^4$	4
11	$\alpha^{11} + 1 = \alpha^2 + \alpha + 1 + 1 = \alpha^{19}$	19
12	$\alpha^{12} + 1 = \alpha^3 + \alpha^2 + \alpha + 1 = \alpha^{23}$	23
13	$\alpha^{13} + 1 = \alpha^4 + \alpha^3 + \alpha^2 + 1 = \alpha^{14}$	14
14	$\alpha^{14} + 1 = \alpha^4 + \alpha^3 + \alpha^2 + 1 + 1 = \alpha^{13}$	13
15	$\alpha^{15} + 1 = \alpha^4 + \alpha^3 + \alpha^2 + \alpha + 1 + 1 = \alpha^{24}$	24
16	$\alpha^{16} + 1 = \alpha^4 + \alpha^3 + \alpha^1 + 1 + 1 = \alpha^9$	9
17	$\alpha^{17} + 1 = \alpha^4 + \alpha^1 + 1 + 1 = \alpha^{30}$	30
18	$\alpha^{18} + 1 = \alpha^1 + 1 + 1 = \alpha^1$	1
19	$\alpha^{19} + 1 = \alpha^2 + \alpha + 1 = \alpha^{11}$	11
20	$\alpha^{20} + 1 = \alpha^3 + \alpha^2 + 1 = \alpha^8$	8
21	$\alpha^{21} + 1 = \alpha^4 + \alpha^3 + 1 = \alpha^{25}$	25
22	$\alpha^{22} + 1 = \alpha^4 + \alpha^2 + 1 + 1 = \alpha^7$	7
23	$\alpha^{23} + 1 = \alpha^3 + \alpha^2 + \alpha + 1 + 1 = \alpha^{12}$	12
24	$\alpha^{24} + 1 = \alpha^4 + \alpha^3 + \alpha^2 + \alpha + 1 = \alpha^{15}$	15
25	$\alpha^{25} + 1 = \alpha^4 + \alpha^3 + 1 + 1 = \alpha^{21}$	21
26	$\alpha^{26} + 1 = \alpha^4 + \alpha^2 + \alpha^1 + 1 + 1 = \alpha^{28}$	28
27	$\alpha^{27} + 1 = \alpha^3 + \alpha^1 + 1 + 1 = \alpha^6$	6
28	$\alpha^{28} + 1 = \alpha^4 + \alpha^2 + \alpha + 1 = \alpha^{26}$	26
29	$\alpha^{29} + 1 = \alpha^3 + 1 + 1 = \alpha^3$	3
30	$\alpha^{30} + 1 = \alpha^4 + \alpha + 1 = \alpha^{17}$	17
31	$\alpha^{31} + 1 = 1 + 1 = 0 = \alpha^{-\infty}$	$-\infty$

Appendix

A1.4 GF(2^6)

If $q = 2$, $m = 6$, then $x^{q^m-1} - 1 = x^{2^6-1} - 1 = x^{63} - 1 = 0$
$x^{63} = 1$

If $f(x) = x^6 + x^1 + 1$, then $f(\alpha) = \alpha^6 + \alpha^1 + 1 = 0$ [α is a root of $f(x)$]
$\alpha^6 = \alpha^1 + 1 \pmod 2$

order is $n = q^6 - 1 = 2^6 - 1 = 63$
If α is a primative element then $\alpha^{63} = 1$ & all elements are powers of α
$\alpha^0, \alpha^1, \alpha^2, \alpha^3, \alpha^4, \alpha^5, \alpha^6, \alpha^7, \alpha^8, \alpha^9, \alpha^{10}, \alpha^{11}, \alpha^{12}, \alpha^{13}, \alpha^{14}, \alpha^{15}, \ldots, \alpha^{62}, \alpha^{63}$

The order of α is $n = 63$ $\rightarrow$ $\alpha^{63} = 1$
Use the multiplication property of the field to form powers of n-tuples
$\alpha^8 \rightarrow 001100$
$(\alpha^{18})^8 = \alpha^{144} = \alpha^{63+63+18} = \alpha^{18} \rightarrow 001111$
$(001100)^8 = 001111$

Mathematics beyond the Calculus

GF(64) - the $m_i(x)$

$$x^{63} - 1 = m_0(x)m_1(x)m_2(x)m_3(x)m_4(x)m_5(x)m_6(x)$$
$$m_7(x)m_8(x)m_9(x)m_{10}(x)m_{11}(x)$$

factor	roots	minimal polynomial	
$x+1$	α^0	$m_0(x)$	0000011
$x^6 + x^1 + 1$	$\alpha^1, \alpha^2, \alpha^4, \alpha^8, \alpha^{16}, \alpha^{32}$	$m_1(x)$	1000011
$x^6 + x^4 + x^2 + x^1 + 1$	$\alpha^3, \alpha^6, \alpha^{12}, \alpha^{24}, \alpha^{48}, \alpha^{33}$	$m_3(x)$	1010111
$x^6 + x^3 + x^2 + x^1 + 1$	$\alpha^5, \alpha^{10}, \alpha^{20}, \alpha^{40}, \alpha^{17}, \alpha^{34}$	$m_5(x)$	1001111
$x^6 + x^3 + 1$	$\alpha^7, \alpha^{14}, \alpha^{28}, \alpha^{56}, \alpha^{49}, \alpha^{35}$	$m_7(x)$	1001001
$x^3 + x^2 + 1$	$\alpha^9, \alpha^{18}, \alpha^{36}$	$m_9(x)$	0001101
$x^6 + x^5 + x^3 + x^2 + 1$	$\alpha^{11}, \alpha^{22}, \alpha^{44}, \alpha^{55}, \alpha^{25}, \alpha^{37}$	$m_{11}(x)$	1101101
$x^6 + x^4 + x^3 + x^1 + 1$	$\alpha^{13}, \alpha^{26}, \alpha^{52}, \alpha^{41}, \alpha^{19}, \alpha^{38}$	$m_{13}(x)$	1011011
$x^6 + x^5 + x^4 + x^2 + 1$	$\alpha^{15}, \alpha^{30}, \alpha^{60}, \alpha^{57}, \alpha^{51}, \alpha^{39}$	$m_{15}(x)$	1110101
$x^2 + x^1 + 1$	α^{21}, α^{42}	$m_{21}(x)$	0000111
$x^6 + x^5 + x^4 + x^1 + 1$	$\alpha^{23}, \alpha^{46}, \alpha^{29}, \alpha^{58}, \alpha^{23}, \alpha^{43}$	$m_{23}(x)$	1110011
$x^3 + x^1 + 1$	$\alpha^{27}, \alpha^{54}, \alpha^{45}$	$m_{27}(x)$	0001011
$x^6 + x^5 + 1$	$\alpha^{31}, \alpha^{62}, \alpha^{61}, \alpha^{59}, \alpha^{55}, \alpha^{54}$	$m_{31}(x)$	1100001

GF(64) $g_i(x)$ (most binaries not shown)

t	g(x)	binary	octal	n	k	r
1	$m_1(x)$	1 000 011	103	63	57	6
2	$m_3(x)m_1(x)$	–	12471	63	51	12
3	$m_5(x)m_3(x)m_1(x)$	–	546217	63	45	18
4	$m_7(x)m_5(x)m_3(x)m_1(x)$	–	126423607	63	39	24
5	$m_{11}(x)m_7(x)m_5(x)m_3(x)m_1(x)$	–	–	63	33	30
6	$m_{13}(x)m_{11}(x)....m_1(x)$	–	–	63	27	36
7	$m_{15}(x)m_{13}(x)....m_1(x)$	–	–	63	21	42
8	$m_{23}(x)m_{15}(x)....m_1(x)$	–	–	63	15	48
9	$m_{31}(x)m_{23}(x)....m_1(x)$	–	–	63	9	54

Appendix

GF(64) elements

elem	polynomial	6tuple	elem	polynomial	6tuple
$\alpha^{-\infty}$	0	000000	α^{33}	$\alpha^4 + \alpha^1$	010010
α^0	1	000001	α^{34}	$\alpha^5 + \alpha^2$	100100
α^1	α	000010	α^{35}	$\alpha^3 + \alpha + 1$	001011
α^2	α^2	000100	α^{36}	$\alpha^4 + \alpha^2 + \alpha^1$	010110
α^3	α^3	001000	α^{37}	$\alpha^5 + \alpha^3 + \alpha^2$	101100
α^4	α^4	010000	α^{38}	$\alpha^4 + \alpha^3 + \alpha^1 + 1$	011011
α^5	α^5	100000	α^{39}	$\alpha^5 + \alpha^4 + \alpha^2 + \alpha^1$	110110
α^6	$\alpha^1 + 1$	000011	α^{40}	$\alpha^5 + \alpha^3 + \alpha^2 + \alpha^1 + 1$	101111
α^7	$\alpha^2 + \alpha^1$	000110	α^{41}	$\alpha^4 + \alpha^3 + \alpha^2 + 1$	011101
α^8	$\alpha^3 + \alpha^2$	001100	α^{42}	$\alpha^5 + \alpha^4 + \alpha^3 + \alpha^1$	111010
α^9	$\alpha^4 + \alpha^3$	011000	α^{43}	$\alpha^5 + \alpha^4 + \alpha^2 + \alpha^1 + 1$	110111
α^{10}	$\alpha^5 + \alpha^4$	110000	α^{44}	$\alpha^5 + \alpha^3 + \alpha^2 + 1$	101101
α^{11}	$\alpha^5 + \alpha + 1$	100011	α^{45}	$\alpha^4 + \alpha^3 + 1$	011001
α^{12}	$\alpha^2 + 1$	000101	α^{46}	$\alpha^5 + \alpha^4 + \alpha^1$	110010
α^{13}	$\alpha^3 + \alpha^1$	001010	α^{47}	$\alpha^5 + \alpha^2 + \alpha^1 + 1$	100111
α^{14}	$\alpha^4 + \alpha^2$	010100	α^{48}	$\alpha^3 + \alpha^2 + 1$	001101
α^{15}	$\alpha^5 + \alpha^3$	101000	α^{49}	$\alpha^4 + \alpha^3 + \alpha^1$	011010
α^{16}	$\alpha^4 + \alpha + 1$	010011	α^{50}	$\alpha^5 + \alpha^4 + \alpha^2$	110100
α^{17}	$\alpha^5 + \alpha^2 + \alpha^1$	100110	α^{51}	$\alpha^5 + \alpha^3 + \alpha^1 + 1$	101011
α^{18}	$\alpha^3 + \alpha^2 + \alpha^1 + 1$	001111	α^{52}	$\alpha^4 + \alpha^2 + 1$	010101
α^{19}	$\alpha^4 + \alpha^3 + \alpha^2 + \alpha^1$	011110	α^{53}	$\alpha^5 + \alpha^3 + \alpha^1$	101010
α^{20}	$\alpha^5 + \alpha^4 + \alpha^3 + \alpha^2$	111100	α^{54}	$\alpha^4 + \alpha^2 + \alpha^1 + 1$	010111
α^{21}	$\alpha^5 + \alpha^4 + \alpha^3 + \alpha^1 + 1$	111011	α^{55}	$\alpha^5 + \alpha^3 + \alpha^2 + \alpha^1$	101110
α^{22}	$\alpha^5 + \alpha^4 + \alpha^2 + 1$	110101	α^{56}	$\alpha^4 + \alpha^3 + \alpha^2 + \alpha^1 + 1$	011111
α^{23}	$\alpha^5 + \alpha^3 + 1$	101001	α^{57}	$\alpha^5 + \alpha^4 + \alpha^3 + \alpha^2 + \alpha^1$	111110
α^{24}	$\alpha^4 + 1$	010001	α^{58}	$\alpha^5 + \alpha^4 + \alpha^3 + \alpha^2 + \alpha^1 + 1$	111111
α^{25}	$\alpha^5 + \alpha^1$	100010			
α^{26}	$\alpha^2 + \alpha^1 + 1$	000111	α^{59}	$\alpha^5 + \alpha^4 + \alpha^3 + \alpha^2 + 1$	111101
α^{27}	$\alpha^3 + \alpha^2 + \alpha^1$	001110	α^{60}	$\alpha^5 + \alpha^4 + \alpha^3 + 1$	111001
α^{28}	$\alpha^4 + \alpha^3 + \alpha^2$	011100	α^{61}	$\alpha^5 + \alpha^4 + 1$	110001
α^{29}	$\alpha^5 + \alpha^4 + \alpha^3$	111000	α^{62}	$\alpha^5 + 1$	100001
α^{30}	$\alpha^5 + \alpha^4 + \alpha^1 + 1$	110011	α^{63}	1	000001
α^{31}	$\alpha^5 + \alpha^2 + 1$	100101			
α^{32}	$\alpha^3 + 1$	001001			

Mathematics beyond the Calculus

GF(64) Z(n)

$$\alpha^{Z(n)} = \alpha^n + 1$$

n	$\alpha^{Z(n)}$	$Z(n)$
0	$\alpha^0 + 1 = 1 + 1 = 0$	$-\infty$
1	$\alpha^1 + 1 = \alpha^6$	6
2	$\alpha^2 + 1 = \alpha^{12}$	12
3	$\alpha^3 + 1 = \alpha^{32}$	32
4	$\alpha^4 + 1 = \alpha^{24}$	24
5	$\alpha^5 + 1 = \alpha^{62}$	62
6	$\alpha^6 + 1 = \alpha + 1 + 1 = \alpha$	1
7	$\alpha^7 + 1 = \alpha^2 + \alpha^1 + 1 = \alpha^{26}$	26
8	$\alpha^8 + 1 = \alpha^3 + \alpha^2 + 1 = \alpha^{48}$	48
9	$\alpha^9 + 1 = \alpha^4 + \alpha^3 + 1 = \alpha^{45}$	45
10	$\alpha^{10} + 1 = \alpha^5 + \alpha^4 + 1 = \alpha^{61}$	61
11	$\alpha^{11} + 1 = \alpha^5 + \alpha + 1 + 1 = \alpha^{25}$	25
12	$\alpha^{12} + 1 = \alpha^2 + 1 + 1 = \alpha^2$	2
13	$\alpha^{13} + 1 = \alpha^3 + \alpha^1 + 1 = \alpha^{35}$	35
14	$\alpha^{14} + 1 = \alpha^4 + \alpha^2 + 1 = \alpha^{52}$	52
15	$\alpha^{15} + 1 = \alpha^5 + \alpha^3 + 1 = \alpha^{23}$	23
16	$\alpha^{16} + 1 = \alpha^4 + \alpha^1 + 1 + 1 = \alpha^{33}$	33
17	$\alpha^{17} + 1 = \alpha^5 + \alpha^2 + \alpha^1 + 1 = \alpha^{47}$	47
18	$\alpha^{18} + 1 = \alpha^3 + \alpha^2 + \alpha^1 + 1 + 1 = \alpha^{27}$	27
19	$\alpha^{19} + 1 = \alpha^4 + \alpha^3 + \alpha^2 + \alpha^1 + 1 = \alpha^{56}$	56
20	$\alpha^{20} + 1 = \alpha^5 + \alpha^4 + \alpha^3 + \alpha^2 + 1 = \alpha^{59}$	59
21	$\alpha^{21} + 1 = \alpha^5 + \alpha^4 + \alpha^3 + \alpha^1 + 1 + 1 = \alpha^{42}$	42
22	$\alpha^{22} + 1 = \alpha^5 + \alpha^4 + \alpha^2 + 1 + 1 = \alpha^{50}$	50
23	$\alpha^{23} + 1 = \alpha^5 + \alpha^3 + 1 + 1 = \alpha^{15}$	15
24	$\alpha^{24} + 1 = \alpha^4 + 1 + 1 = \alpha^4$	4
25	$\alpha^{25} + 1 = \alpha^5 + \alpha^1 + 1 = \alpha^{11}$	11
26	$\alpha^{26} + 1 = \alpha^2 + \alpha^1 + 1 + 1 = \alpha^7$	7
27	$\alpha^{27} + 1 = \alpha^3 + \alpha^2 + \alpha^1 + 1 = \alpha^{18}$	18
28	$\alpha^{28} + 1 = \alpha^4 + \alpha^3 + \alpha^2 + 1 = \alpha^{41}$	41
29	$\alpha^{29} + 1 = \alpha^5 + \alpha^4 + \alpha^3 + 1 = \alpha^{60}$	60
30	$\alpha^{30} + 1 = \alpha^5 + \alpha^4 + \alpha^1 + 1 + 1 = \alpha^{46}$	46
31	$\alpha^{31} + 1 = \alpha^5 + \alpha^2 + 1 + 1 = \alpha^{34}$	34
32	$\alpha^{32} + 1 = \alpha^3 + 1 + 1 = \alpha^3$	3

Appendix

GF(64) Z(n) continued

n	$\alpha^{Z(n)}$	$Z(n)$
33	$\alpha^{33}+1 = \alpha^4+\alpha^1+1 = \alpha^{16}$	16
34	$\alpha^{34}+1 = \alpha^5+\alpha^2+1 = \alpha^{18}$	31
35	$\alpha^{35}+1 = \alpha^3+\alpha^1+1+1 = \alpha^{13}$	13
36	$\alpha^{36}+1 = \alpha^4+\alpha^2+\alpha^1+1 = \alpha^{54}$	54
37	$\alpha^{37}+1 = \alpha^5+\alpha^3+\alpha^2+1 = \alpha^{44}$	44
38	$\alpha^{38}+1 = \alpha^4+\alpha^3+\alpha^1+1+1 = \alpha^{49}$	49
39	$\alpha^{39}+1 = \alpha^5+\alpha^4+\alpha^2+\alpha^1+1 = \alpha^{43}$	43
40	$\alpha^{40}+1 = \alpha^5+\alpha^3+\alpha^2+\alpha^1+1+1 = \alpha^{55}$	55
41	$\alpha^{41}+1 = \alpha^4+\alpha^3+\alpha^2+1+1 = \alpha^{28}$	28
42	$\alpha^{42}+1 = \alpha^5+\alpha^4+\alpha^3+\alpha^1+1 = \alpha^{21}$	21
43	$\alpha^{43}+1 = \alpha^5+\alpha^4+\alpha^2+\alpha^1+1+1 = \alpha^{39}$	39
44	$\alpha^{44}+1 = \alpha^5+\alpha^3+\alpha^2+1+1 = \alpha^{37}$	37
45	$\alpha^{45}+1 = \alpha^4+\alpha^3+1+1 = \alpha^9$	9
46	$\alpha^{46}+1 = \alpha^5+\alpha^4+\alpha^1+1 = \alpha^{30}$	30
47	$\alpha^{47}+1 = \alpha^5+\alpha^2+\alpha^1+1+1 = \alpha^{17}$	17
48	$\alpha^{48}+1 = \alpha^3+\alpha^2+1+1 = \alpha^8$	8
49	$\alpha^{49}+1 = \alpha^4+\alpha^3+\alpha^1+1 = \alpha^{38}$	38
50	$\alpha^{50}+1 = \alpha^5+\alpha^4+\alpha^2+1 = \alpha^{22}$	22
51	$\alpha^{51}+1 = \alpha^5+\alpha^3+\alpha^1+1+1 = \alpha^{53}$	53
52	$\alpha^{52}+1 = \alpha^4+\alpha^2+1+1 = \alpha^{14}$	14
53	$\alpha^{53}+1 = \alpha^5+\alpha^3+\alpha^1+1 = \alpha^{51}$	51
54	$\alpha^{54}+1 = \alpha^4+\alpha^2+\alpha^1+1+1 = \alpha^{36}$	36
55	$\alpha^{55}+1 = \alpha^5+\alpha^3+\alpha^2+\alpha^1+1 = \alpha^{40}$	40
56	$\alpha^{56}+1 = \alpha^4+\alpha^3+\alpha^2+\alpha^1+1+1 = \alpha^{19}$	19
57	$\alpha^{57}+1 = \alpha^5+\alpha^4+\alpha^3+\alpha^2+\alpha^1+1 = \alpha^{58}$	58
58	$\alpha^{58}+1 = \alpha^5+\alpha^4+\alpha^3+\alpha^2+\alpha^1+1+1 = \alpha^{57}$	57
59	$\alpha^{59}+1 = \alpha^5+\alpha^4+\alpha^3+\alpha^2+1+1 = \alpha^{20}$	20
60	$\alpha^{60}+1 = \alpha^5+\alpha^4+\alpha^3+1+1 = \alpha^{29}$	29
61	$\alpha^{61}+1 = \alpha^5+\alpha^4+1+1 = \alpha^{10}$	10
62	$\alpha^{62}+1 = \alpha^5+1+1 = \alpha^5$	5
63	$\alpha^{63}+1 = 1+1 = 0$	0

Mathematics beyond the Calculus

Galois Algebra Review

The French mathematician Evariste Galois created the algebra of finite fields. A finite field of q elements is denoted as GF(q) where q is a prime number. The number of elements in an extension of GF(q) must be q^m, where q is a prime number and m is a positive integer. The extension field is written as GF(q^m). If q=2, then the elements of GF(2) are 0 and 1. In this text q=2.

Every extension field has a primitive element α that generates the 2^m-1 elements of GF(2^m). For example when m=4 the order and elements are

(101a) *order is* $q^4 - 1 = 2^4 - 1 = 15$

If α is a primative element then $\alpha^{15} = 1$ *and all elements are powers of α*
(101b) $\alpha^0, \alpha^1, \alpha^2, \alpha^3, \alpha^4, \alpha^5, \alpha^6, \alpha^7, \alpha^8, \alpha^9, \alpha^{10}, \alpha^{11}, \alpha^{12}, \alpha^{13}, \alpha^{14}$

As written, the 15 elements cannot be represented by 4 bit binary numbers such as 1011. E.g. α^7=1000000. Here is how exponents of elements greater than 3 are reduced.

Fermat's equation defines the *order* of α as 2^m-1.

(102) $\alpha^{2^m-1} = 1 \;\Rightarrow\; \alpha^{2^m-1} + 1 = 1 + 1 \;\Rightarrow\; \alpha^{2^m-1} + 1 = 0$

Every element of field GF(2^m) is shown in (101b). *One of the factors of the Fermat equation is the polynomial that generates the field elements.*

(103) $\alpha^{2^m-1} = 1 \;\Rightarrow\; \alpha^{2^4-1} = \alpha^{15} = 1 \;\Rightarrow\; \alpha^{15} + 1 = 0$

Divisors of $d = 15$ are 1, 3, 5 and 15, use 1
$x^{15} + 1 = (x+1)(x^{14} + \cdots + 1)$
factor x^{14}.... Divisors of $d = 14$ are 2, 7, use 2
$x^{15} + 1 = (x+1)(x^2 + x + 1)(x^{12} + \cdots + 1)$ (mod 2)
factor x^{12}.... Divisors of $d = 12$ are 2, 3, 4, 6 use 4
(104) $x^{15} + 1 = (x+1)(x^2 + x + 1)(x^4 + x^3 + x^2 + x + 1)(x^4 + x^3 + 1)(x^4 + x + 1)$

Appendix

One way to reduce polynomials to degree 2^m-1 or less is to use the fact α is a root of one of the factors $x^{15}+1=0$. We can select any factor such as polynomial $f(x)=x^4+x+1$ (from equation 104).

If $f(x) = x^4 + x + 1$, then $f(\alpha) = \alpha^4 + \alpha + 1 = 0$ and so

(105) $\alpha^4 = \alpha + 1$

We use this equation for α^4 to reduce powers of α to polynomials of degree 3 or less.

GF(16) elements

element	polynomial	4tuple	hex
$\alpha^{-\infty}$	0	0000	0
α^0	1	0001	1
α^1	α	0010	2
α^2	α^2	0100	4
α^3	α^3	1000	8
α^4	$\alpha+1$	0011	3
α^5	$\alpha^2+\alpha$	0110	6
α^6	$\alpha^3+\alpha^2$	1100	C
α^7	$\alpha^3+\alpha+1$	1011	B
α^8	α^2+1	0101	5
α^9	$\alpha^3+\alpha$	1010	A
α^{10}	$\alpha^2+\alpha+1$	0111	7
α^{11}	$\alpha^3+\alpha^2+\alpha$	1110	E
α^{12}	$\alpha^3+\alpha^2+\alpha+1$	1111	F
α^{13}	$\alpha^3+\alpha^2+1$	1101	D
α^{14}	α^3+1	1001	9
α^{15}	1	0001	1

Multiplication Now we have a straightforward way to find y^3. For example

(106a) If $y = \alpha^2$, then $y^3 = (\alpha^2)^3 = \alpha^6$

(106b) If $y = \alpha^{11}$, then $y^3 = (\alpha^{11})^3 = \alpha^{33} = \alpha^{15}\alpha^{15}\alpha^3 = 1\cdot 1\cdot \alpha^3 = \alpha^3$

Addition Addition in $GF(2^4)$ is straightforward. The polynomial form of the elements is used in the addition operation. Examples are

(107a) $\alpha^7 + \alpha^2 = (\alpha^3 + \alpha + 1) + \alpha^2 = \alpha^{12}$

(107b) $\alpha^4 + \alpha^5 = (\alpha+1) + (\alpha^2+\alpha) = \alpha^2 + 1 = \alpha^8$

(107c) $\alpha^5 + \alpha^{13} = (\alpha^2+\alpha) + (\alpha^3+\alpha^2+1) = \alpha^3 + \alpha + 1 = \alpha^7$

Mathematics beyond the Calculus

Answers to Most of the Problems

Problems
(101) $(a-b)(b-c)(c-a)(a+b+c)$
(102) $(x-y)(x-z)(y-z)(x+y+z)$
(103) $(x-y)(x-z)(y-z)(x^2+y^2+z^2+xy+xz+yz)$
(104) 4abc (105) $(a-b)^3$
(106) 84 (107) –5 (108) –36
(109) 29 (110) –15 (111) –27
(112) $-(\lambda-2)(\lambda+1)^2$ (113) $(\lambda-1)^3$
(114) x= –6, 1
(115) x=6/7 y=10/7 z=5/7

Problem 201

$$AB = \begin{bmatrix} 5 & 18 \\ 14 & 39 \end{bmatrix} \quad BA = \begin{bmatrix} 1 & 2 & 3 \\ 14 & 19 & 24 \\ 16 & 20 & 24 \end{bmatrix}$$

Problem 202

If $AC = CA$ and $BC = CB$ where $C = \begin{bmatrix} 0 & 1 \\ -1 & 0 \end{bmatrix}$ then show that $AB = BA$

$$AC = \begin{bmatrix} p & q \\ r & s \end{bmatrix}\begin{bmatrix} 0 & 1 \\ -1 & 0 \end{bmatrix} = \begin{bmatrix} -q & p \\ -s & r \end{bmatrix} \quad \text{and} \quad CA = \begin{bmatrix} 0 & 1 \\ -1 & 0 \end{bmatrix}\begin{bmatrix} p & q \\ r & s \end{bmatrix} = \begin{bmatrix} r & s \\ -p & -q \end{bmatrix}$$

if $AC = CA$ then $\begin{bmatrix} -q & p \\ -s & r \end{bmatrix} = \begin{bmatrix} r & s \\ -p & -q \end{bmatrix}$ so that $-q = r \quad p = s \quad A = \begin{bmatrix} p & q \\ -q & p \end{bmatrix}$

similiarly $B = \begin{bmatrix} t & u \\ -u & t \end{bmatrix} \quad AB = \begin{bmatrix} p & q \\ -q & p \end{bmatrix}\begin{bmatrix} t & u \\ -u & t \end{bmatrix} = \begin{bmatrix} pt-qu & pu+qt \\ -qt-pu & -qu+pt \end{bmatrix}$

$BA = \begin{bmatrix} t & u \\ -u & t \end{bmatrix}\begin{bmatrix} p & q \\ -q & p \end{bmatrix} = \begin{bmatrix} tp-uq & tq+up \\ -up-tq & -uq+tp \end{bmatrix}$ therefore $AB = BA$

Problem 203 Write the matrix equation corresponding to these equations.
$$12 = 2x_1 + 5x_2 + 7x_3$$
$$0 = 3x_1 - 6x_2 + x_3$$
$$5 = -x_1 + 7x_2 + 3x_3$$

Answers to most of the Problems

Problem 204

$$\begin{bmatrix} \dfrac{d_{22}d_{33}}{d_{11}d_{22}d_{33}} & 0 & 0 \\ 0 & \dfrac{d_{11}d_{33}}{d_{11}d_{22}d_{33}} & 0 \\ 0 & 0 & \dfrac{d_{11}d_{22}}{d_{11}d_{22}d_{33}} \end{bmatrix} = \begin{bmatrix} \dfrac{1}{d_{11}} & 0 & 0 \\ 0 & \dfrac{1}{d_{22}} & 0 \\ 0 & 0 & \dfrac{1}{d_{33}} \end{bmatrix}$$

Problem 205

$A^3 - 4A^2 + A + I = 0 \;\rightarrow\; -(A^3 - 4A^2 + A) = I$
$\rightarrow \; -(A^2 - 4A^1 + I)A = I = A^{-1}A$
$(A^2 - 4A^1 + I) = A^{-1}$

Problem 206

solve for A when $(A^T B)^{-1} - (B^T A^{-1})^{-1} + (B^{-1} A^T)^T = I$
use reverse rules $B^{-1}(A^T)^{-1} - A(B^T)^{-1} + A(B^{-1})^T = I$
invoke $(B^T)^{-1} = (B^{-1})^T \;\rightarrow\; B^{-1}(A^T)^{-1} - A(B^{-1})^T + A(B^{-1})^T = I$
cancel terms $B^{-1}(A^T)^{-1} = I \;\rightarrow\;$ let $(A^T)^{-1} = B$ so that $B^{-1}B = I$
then $(A^T)^{-1} = B \;\rightarrow\; A^T = B^{-1} \;\rightarrow\; A = (B^{-1})^T$

Problem 207

matrices C and D commute if $CD = DC$
if A^T and B^T commute then
$A^T B^T = B^T A^T \;\rightarrow\; (A^T B^T)^T = (B^T A^T)^T \;\rightarrow\; BA = AB$

Mathematics beyond the Calculus

Problem 301

$\lambda^3 - 4\lambda^2 - 3\lambda + 18 = 0 \rightarrow (\lambda+3)^2(\lambda+2) = 0 \rightarrow x_1 = \begin{bmatrix} -1 \\ 0 \\ 2 \end{bmatrix} \quad x_2 = \begin{bmatrix} -1 \\ 2 \\ 1 \end{bmatrix}$

Problem 302

$Ax = \lambda x \rightarrow (A - \lambda I)x = 0 \quad \begin{bmatrix} 3-\lambda & 6 & -8 \\ 0 & 0-\lambda & 6 \\ 0 & 0 & 2-\lambda \end{bmatrix} \begin{bmatrix} x_1 \\ x_2 \\ x_3 \end{bmatrix} = 0$

$\det(A - \lambda I) = (3-\lambda)(0-\lambda)(2-\lambda)$

$\begin{bmatrix} (3-\lambda)x_1 + 6x_2 - 8x_3 \\ -\lambda x_2 + 6x_3 \\ (2-\lambda)x_3 \end{bmatrix} = 0$

for $\lambda = 2 \quad Ax - 2x = \begin{bmatrix} (3-2)x_1 + 6x_2 - 8x_3 \\ -2x_2 + 6x_3 \\ (2-2)x_3 \end{bmatrix} = \begin{bmatrix} x_1 + 6x_2 - 8x_3 \\ -2x_2 + 6x_3 \\ 0 \end{bmatrix} = 0 \rightarrow x = \begin{bmatrix} -10 \\ 3 \\ 1 \end{bmatrix}$

for $\lambda = 0 \quad Ax = \begin{bmatrix} 3x_1 + 6x_2 - 8x_3 \\ 6x_3 \\ 2x_3 \end{bmatrix} = 0 \rightarrow x = \begin{bmatrix} -2 \\ 1 \\ 0 \end{bmatrix}$

for $\lambda = 3 \quad Ax - 3x = \begin{bmatrix} (3-2)x_1 + 6x_2 - 8x_3 \\ -3x_2 + 6x_3 \\ (2-3)x_3 \end{bmatrix} = \begin{bmatrix} x_1 + 6x_2 - 8x_3 \\ -3x_2 + 6x_3 \\ -x_3 \end{bmatrix} = 0 \rightarrow x = \begin{bmatrix} 0 \\ 0 \\ 0 \end{bmatrix}$

Problem 1415

if $\mathcal{L}[y] = \dfrac{1}{p(p^2 - \omega^2)}$ *show that* $y = \mathcal{L}^{-1}[y] = -\dfrac{1}{\omega^2} + \dfrac{1}{2\omega^2}e^{-\omega t} + \dfrac{1}{2\omega^2}e^{\omega t}$

$y = -\dfrac{1}{\omega^2} + \dfrac{1}{\omega^2}\cosh \omega t = \dfrac{1}{\omega^2}(\cosh \omega t - 1)$

Problem 1416

show that $y = \dfrac{1}{\omega^2}(\cosh \omega t - 1)$

is a solution of $\dfrac{d^2 y}{dt^2} - \omega^2 y = 1$ *when* $y(0) = y'(0) = 0$

Answers to most of the Problems

Problem 1417

if $\mathscr{L}[i] = \dfrac{Li(0)}{(R+pL)} + \dfrac{V_m}{(R+pL)(p-j\omega)}$

show that $\mathscr{L}^{-1}[i] = i(0)e^{-\frac{R}{L}t} + \dfrac{V_m}{R+j\omega L}e^{j\omega t} - \dfrac{V_m}{R+j\omega L}e^{-\frac{R}{L}t}$

(22) $I(p) = \dfrac{1}{L} \cdot \dfrac{1}{p+\frac{R}{L}} \cdot \dfrac{V_m}{p-j\omega} + \dfrac{i(0)}{p+\frac{R}{L}}$

$I(p) = \dfrac{1}{L} \cdot \dfrac{1}{j\omega+\frac{R}{L}} \cdot \dfrac{V_m}{p-j\omega} + \dfrac{1}{L} \cdot \dfrac{1}{p+\frac{R}{L}} \cdot \dfrac{V_m}{-\frac{R}{L}-j\omega} + \dfrac{i(0)}{p+\frac{R}{L}}$

$I(p) = \dfrac{V_m}{R+j\omega L} \cdot \dfrac{1}{p-j\omega} - \dfrac{V_m}{R+j\omega L} \cdot \dfrac{1}{p+\frac{R}{L}} + \dfrac{V_b}{R} \cdot \dfrac{1}{p+\frac{R}{L}}$

Problem 1418

show that $i = \dfrac{V_m}{R+j\omega L}e^{j\omega t}$ or $i = -\dfrac{V_m}{R+j\omega L}e^{-\frac{R}{L}t}$ or $i = i(0)e^{-\frac{R}{L}t}$

are solutions of $L\dfrac{di}{dt} + Ri = V_m e^{j\omega t}$ when $i(0) = \dfrac{V_m}{R}$

Problem 1419 Reference example 3.
show that $y = y(0) + y'(0)(1-e^{-t})$

Index

Appendix 118
Galois Field Equations 118
GF(2^3) 119
GF(2^4) 121
GF(2^5) 124
GF(2^6) 128
Galois Algebra Review 133

Determinants 1
Cramer's Rule 5
Evaluation of Determinants 3
Expansion of Determinants 4
Minors and Cofactors 4
Properties of Determinants 2
Solution of Equations 1

Difference Equations 100
Elementary Sequences 100
Solution by Z Transform 101

Eigenvalues 19
Eigenvectors 19

Finite Matrices 7
Cancel Common Factors 15
Complex Elements 15
Definition 10
Inverse of a Square Matrix 14
Inverse Reversal Rule 15
Linear Dependence 16
Matrix Multiplication 11
Matrix Sums and Differences .. 13
Rank of a Matrix 16
Scalar Multiplication 16
The Adjoint Matrix 13
The (7,4) Decoder 7
Transpose of a Matrix 16
Vandermonde Matrices 17

Fourier Series 85

Functions of a Complex
Variable 44
Complex Numbers 44
Analytic Functions 48
Integration 50
Closed contours 50
Line integrals 51
Cauchy's Theorem 52
Cauchy's Integral formula 54
Singularities 55
Laurent Power Series 55
Residues 57
Higher order poles 58
Residue Theorem 59
Evaluation of integrals 60
Inverse Laplace Transform 62

Galois Finite Fields GF(2^m) 103
Sys G Matrix Generator g(x) ... 115
Irreducible Polynomials 110
Minimal Polynomials 111
Polynomial Operations 107
Primitive Roots and Polymials 113
Systematic G and H Matrices .. 116
Zech Logarithms 114

Laplace Transform 21
Gamma Function 34
General Transforms 23
Laplace Transform 21
More Properties of Transforms 35
Partial Fractions 29
Periodic Functions 33
Prologue 22
Specific Transforms 26
Tables of Transforms 41

Ordinary Diff Equations 65
Solution by Laplace Transform 65
Solution by Different Operator 68
Complementary solution y_c 68
Particular solution y_p 70
Reduction of order method 70
Undetermined coeff method 71

Index

Partial Differential Equations 78
Soln by Separation of Variables 78
Initial and Boundary conditions 80
Solution by Laplace Transform . 81

Systems of Ordinary
Differential Equations .74
Solutn by Differential Operator . 74
Solution by Laplace Transform . 77

Z Transform 89
General Z Transforms 91
Inverse Z Transforms 95
Specific Z Transforms 93
Tables of Transforms 98
Z Transform Defined 89

www.ingramcontent.com/pod-product-compliance
Lightning Source LLC
Chambersburg PA
CBHW052320220526
45472CB00001B/197